GCSE MATHEMATICS

LONGMAN
REFERENCE
GUIDES

Angie Cross

LONGMAN REFERENCE GUIDES

Series editors: Geoff Black and Stuart Wall

TITLES AVAILABLE
CDT
English
French
Mathematics
Science
World History

FORTHCOMING
Biology
Chemistry
Geography
Physics

MATHEMATICS

LONGMAN REFERENCE GUIDES

Brian Speed

Longman

Longman Group UK Limited,
Longman House, Burnt Mill, Harlow,
Essex CM20 2JE, England
and Associated Companies throughout the world.

First published 1989

British Library Cataloguing in Publication Data

Speed, Brian
Mathematics.
1. England. Secondary schools. Curriculum subjects:
Mathematics. G.C.S.E. examinations
I. Title
510'.76

ISBN 0-582-05074-X

Designed and produced by The Pen and Ink Book Company Ltd, Huntingdon, Cambridgeshire

Illustrated by Chris Etheridge

Set in 9/10pt Century Old Style

Printed and bound in Great Britain

HOW TO USE THIS BOOK

Throughout your GCSE course you will be coming across terms, ideas and definitions that are unfamiliar to you. The Longman Reference Guides provide a quick, easy-to-use source of information, fact and opinion. Each main term is listed alphabetically and, where appropriate, cross-referenced to related terms.

- Where a term or phrase appears in **different type** you can look up a separate entry under that heading elsewhere in the book.
- Where a term or phrase appears in **different type** and is set between two arrowhead symbols ◀ ▶, it is particularly recommended that you turn to the entry for that heading.

ACKNOWLEDGEMENTS

I am indebted to the following Examination Groups for permission to use some of their GCSE questions in this book: London and East Anglian Group (LEAG); Midlands Examining Group (MEG); Northern Examining Association (NEA); Northern Ireland Schools Examinations Council (NISEC); Southern Examining Group (SEG).

INTRODUCTION

This reference guide is all about GCSE mathematics, so the definitions and the uses are those common to GCSE. In some cases the full mathematical implication of a word or term would be *beyond* the requirements of GCSE. Where this is so we *restrict* our discussion to the common use of that term in the context of GCSE mathematics.

The reference guide is in alphabetical order, which should make it easier for you to look up all the details of a particular term or topic in mathematics. Each entry will define the term you have looked up and give you any associated words or linked expressions to help you understand that entry more fully. Within each entry are comments on the common mistakes and errors that students frequently make during an examination. Not every entry will have these mistakes highlighted; only those for which a particular error is very commonly made.

There are some examples and examination questions in most entries to help you see the kind of question that will be asked using that term or on that topic. I have labelled each question with the appropriate level of question, that is:

H for Higher Level only
I for Intermediate Level
B for Basic level.

Do not forget that even if you are at the Higher level of entry, you still need to be aware of questions set at lower levels, and the Intermediate level student needs to be aware of the Basic level questions.

This book is not written with the intention of being read from cover to cover, but should be delved into at the point at which you have a problem or require further information. So you can have this book available as a constant course companion throughout the last two years of your GCSE course, helping with the coursework, homework and examination preparation.

ACCELERATION

This is the rate of change of **speed** or **velocity** with respect to time. The standard unit is metres per second per second (m/s^2). The *gradient* on a velocity/time graph will represent the acceleration.

ACCURACY

Accuracy often depends on **rounding** off an answer to the current number of **significant figures** or **decimal places** (Fig A.1). You could be asked to round off to the nearest penny or cm, etc., in which case you round off to the nearest *whole number*. You will find that some questions will ask you to round off to a *particular degree* of accuracy, while others do not.

	Significant figures		
Number	*1*	*2*	*3*
13947	10000	14000	13900
46.85	50	47	46.9
0.004193	0.004	0.0042	0.00419

	Decimal places		
Number	*1*	*2*	*3*
26.4791	26.5	26.48	26.479
0.0815	0.1	0.08	0.082
5.1972	5.2	5.20	5.197

Fig A.1

If the question *does* ask for a particular degree of accuracy, then simply do just that. But, if the question does *not* state anything about accuracy, then you have to round off to what you decide is a *sensible* answer. If you are in doubt, then a good guide is to round off to *one more significant figure* than the figures used in the question. For example:

- Find the area of the rectangle 5.6 cm by 9.7 cm.

Since the numbers given in the question are both to *two* significant figures, round off to *three significant figures*. In this case your answer should be 54.3 cm.

Of course there are some situations in which common sense should tell you that the answer must be a whole number:

- How many days will a tin of cocoa last?
- How many sheep can you get in a particular size pen?

Never round off *too soon*, especially when there are a number of stages to be worked out, as this will usually result in an inaccurate final answer.

Always *check* your answer to see if it is sensible and that it is rounded off to a suitable degree of accuracy. Questions will not usually examine rounding off as a topic in itself, but it will be included in quite a few questions throughout your examination. Unless you round off correctly in these questions you will lose marks.

Remember too that a safeguard against being inaccurate is to show all the stages in your *method of solution*. You will then get the majority of marks even if you make an arithmetical mistake in your calculation.

◀ Rounding off ▶

ADDING/ADDITION

◀ Directed numbers, Matrices ▶

ADJACENT

Adjacent means 'next to'. In the case of a right-angled triangle we are often concerned with finding the *adjacent side*. This is often easiest to find by *eliminating* two of the three sides.

- Eliminate the **hypotenuse** – the side opposite the right angle.
- Eliminate the *side opposite* the angle in question.
- The remaining side is the *adjacent side*.

In Figure A.2a), the angle in question is $C\hat{A}B$. The *side adjacent* to $C\hat{A}B$ is therefore AC. In Fig A.2b), the angle in question is $A\hat{C}B$. The *side adjacent* to $A\hat{C}B$ is therefore AC. Notice how the adjacent side is the one next to *both* the right angle and the angle of the question.

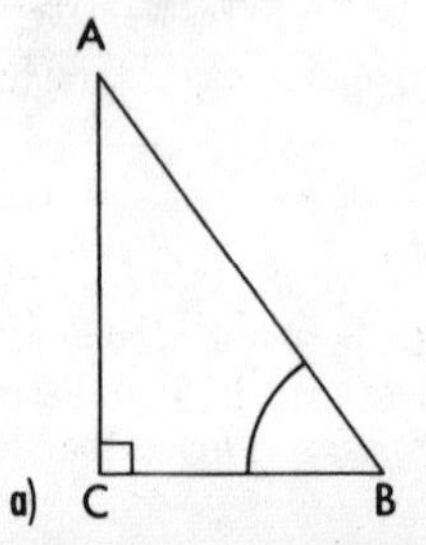

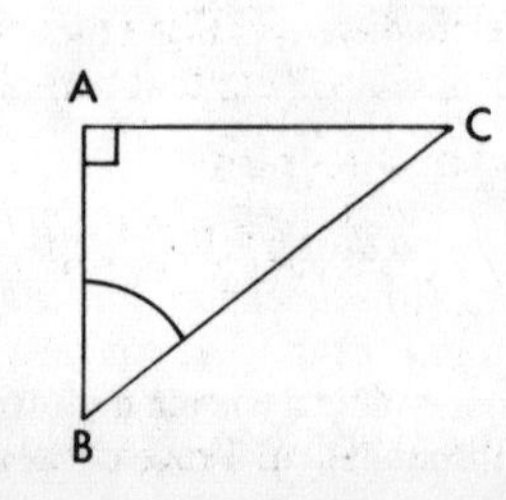

Fig A.2 a) b)

ALGEBRA

Algebra is the part of mathematics that uses letters for numbers. It is a large topic that includes **solving** equations, **factorising, transforming equations, generalising,** using **formulae** and **substitution.** Algebra is used at all levels of GCSE mathematics; simple algebra at the Basic Level and more complicated algebra at the Higher Level.

One of the most important things in GCSE Algebra questions is to recognise that the question *is* about algebra. Many questions now use the *words* rather than the letters, so you need to *recognise* that expressing the information given in the form of a simple equation will help solve the problem. Take the following question at Basic Level:

- **Exam Question**

  In his will Ken leaves all his estate to be shared among three people. Alice is to have £3000 more than Brian. Cyril is to have four times as much as Brian.

  a) If Cyril received £8000, how much would Alice get?

  b) When Ken died, Alice and Cyril received the same amount of money. How much was Ken's estate?

- **Solution**

  This is an algebra-type question that can best be solved by using simple equations. For instance:

  If Brian receives £x

  Alice will get x + 3000

  Cyril will get 4x

  a) If 4x = 8000

  then x = 2000,

  hence Alice gets 2000 + 3000 = £5000.

  b) Where 4x = x + 3000

  3x = 3000

  x = 1000

  This means that Alice received £4000, Brian £1000 and Cyril £4000. So altogether they received £9000, the value of Ken's estate.

The major mistake students made with this question was in *not recognising* that it was an algebra-type question and trying to solve it by trial and error rather than using simple equations. Trial and error takes longer, often leads to wrong answers and scores few marks for method.

The Intermediate and Higher level GCSE examination papers will have quite a few questions in them that use algebra:

- **Exam Question**

  The cost, C pence, of making a clown's hat in the shape of the cone in Figure A.3 with radius r cm and slant height l cm, is given by the formula:

  $$C = 2\pi r + \pi r l,$$

  and rounded off to the nearest penny.

a) Fully factorise the right-hand side of the equation.
b) Hence, or otherwise, transform the formula to make r the subject.
c) Hence, or otherwise, find the radius of a clown's hat with a slant height of 18 cm and costing £4.40.

(NEA; I)

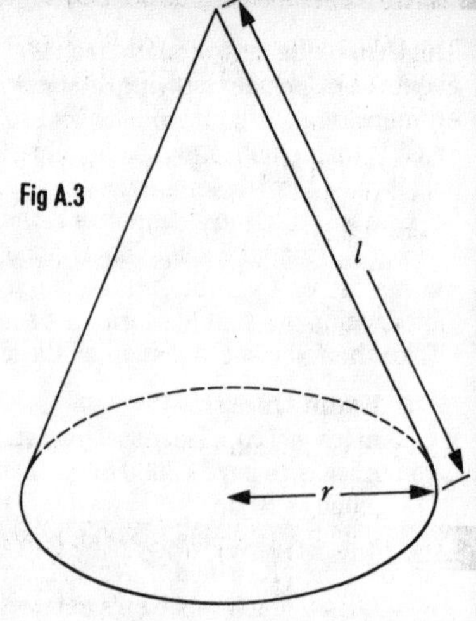

Fig A.3

■ **Solution**

a) $C = \pi r\ (2 + l)$

b) $r = \dfrac{C}{\pi\ (2 + l)}$

c) when $C = 440$, $l = 18$, then $r = \dfrac{440}{\pi\ (2 + 18)} = 7.0028$.

A sensible rounding is 7 cm.

ALGEBRAIC FRACTIONS

These are **fractions** that have *letters* in them, e.g. $\frac{3}{x}$, $\frac{y + 3}{5 - y}$

They will only be found in the higher levels of mathematics and usually in the form of equations to solve. For example:

$$\frac{3}{x} + \frac{2}{x+1} = 1$$

$$\frac{3\ (x + 1) + 2x}{x\ (x + 1)} = 1 \quad \text{get a common denominator}$$

$$3\ (x + 1) + 2x = x\ (x + 1) \quad \text{cross multiply}$$
$$5x + 3 = x^2 + x \quad \text{expand}$$
$$x^2 - 4x - 3 = 0 \quad \text{simplify}$$

You would now solve this quadratic equation.

ALLIED ANGLES

◀ Transversal ▶

ALTERNATE ANGLES

◀ Transversal ▶

'AND' RULE

The 'AND' rule comes from the topic of **probability**. To find the probability of event A **and** event B, the probability of each event is *multiplied* together. For example:

A draw contains five green socks and four yellow socks. What is the probability of picking out a pair of green socks one dark night when you can't see the colours properly?

The probability of the first sock being green is $\frac{5}{9}$
The probability of the second sock being green is $\frac{4}{8}$
Hence the probability of both the first **and** the second sock being green is

$$\frac{5}{9} \times \frac{4}{8} = \frac{20}{72} = \frac{5}{18}$$

The most common error to be made in this is to add the fractions instead of *multiplying* them.

ANGLE

Angle is the amount of turn, and is measured in degrees.

Types of angle

- A *right* angle is a quarter turn, or 90°.
- An *acute* angle is less than 90°.
- An *obtuse* angle is greater than 90°, but less than 180°.
- A *reflex* angle is greater than 180°, but less than 360°.

Facts about angles

1 Angles on a straight line add up to 180° (Fig A.4).

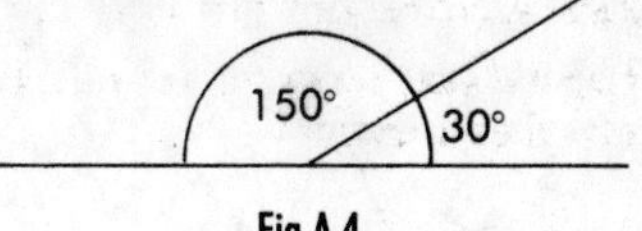

Fig A.4

2 Vertically opposite angles are equal (Fig A.5).

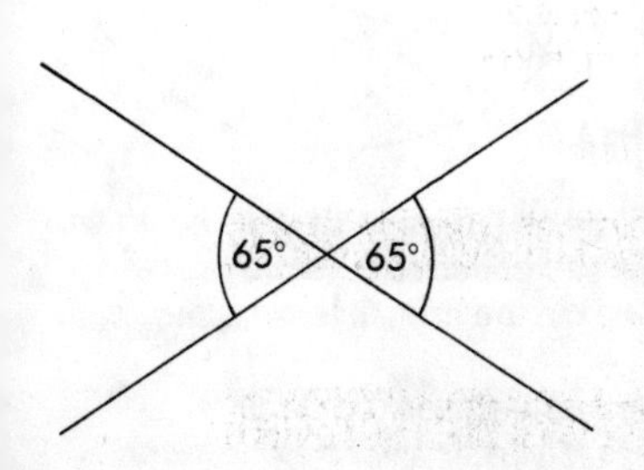

Fig A.5

3 Angles around a point add up to 360° (Fig A.6).

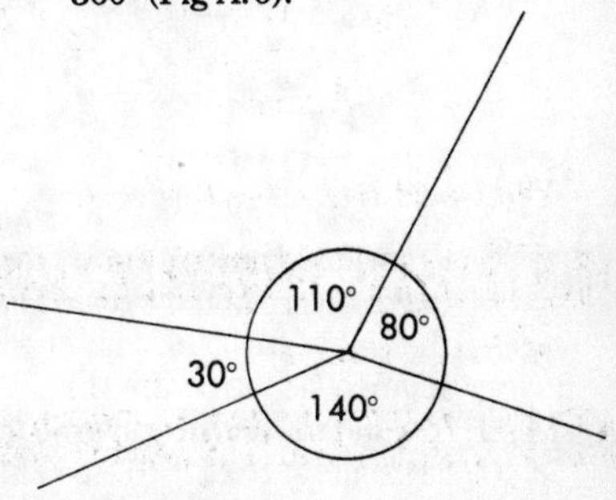

Fig A.6

4 The three angles in a triangle add up to 180° (Fig A.7).

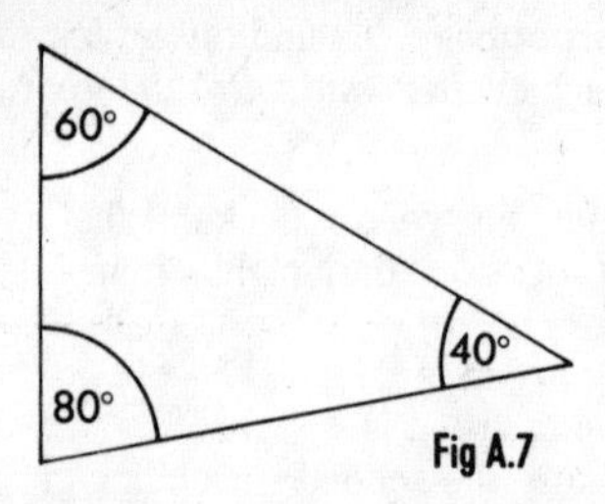

Fig A.7

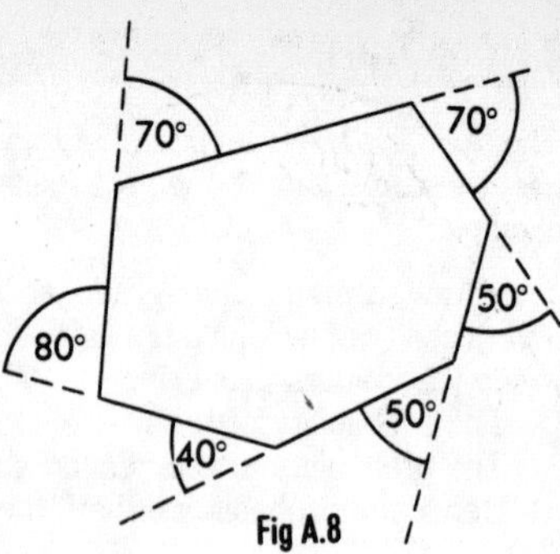

Fig A.8

5 The 'N' angles inside an 'N'-sided polygon will *add up* to 180° (N−2)°. These angles are called the *interior angles.*

6 The 'N' *exterior angles* of an 'N'-sided polygon will *add up to* 360° (Fig A.8).

7 The size of *each interior angle* in an 'N'-sided polygon is given by $180° - \frac{360°}{N}$

8 The size of *each exterior angle* in an 'N'-sided polygon is given by $\frac{360°}{N}$

9 From any **chord** in a circle there are many *triangles* that can be formed in the same segment that touch the circumference, as shown in Figure A.9. All these angles from the same chord are *equal.*

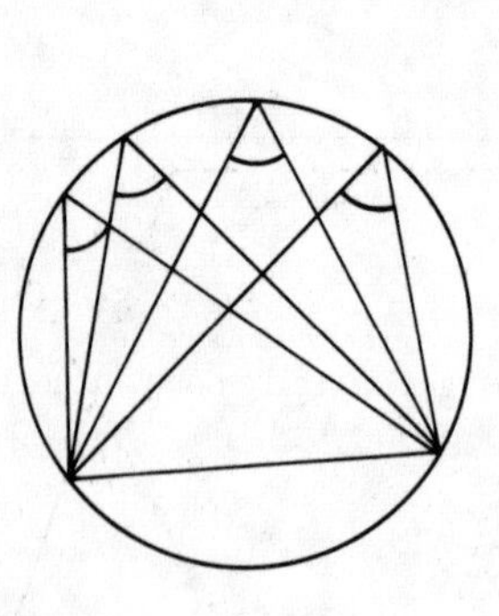
Fig A.9

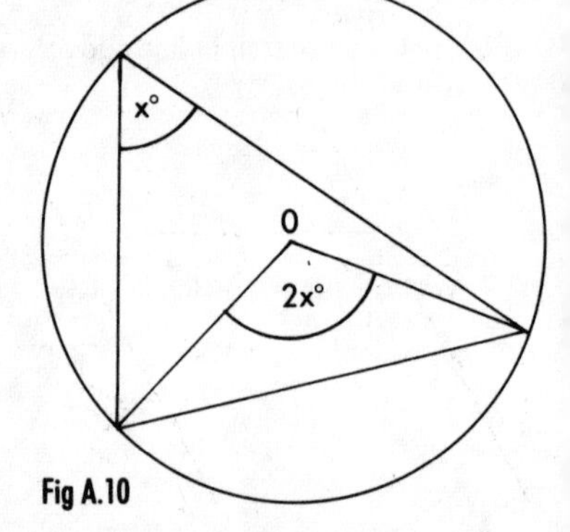

Fig A.10

10 From any *chord* in a circle, there is only one *triangle* that can be drawn to the centre of the circle. The angle subtended at the centre is twice the size of any angle subtended at the circumference in the same segment as the centre (Fig A.10).

11 Any **quadrilateral** drawn so that the four vertices touch the circumference of the same circle is called **cyclic** and its *opposite angles* add up to 180° (Fig A.11).

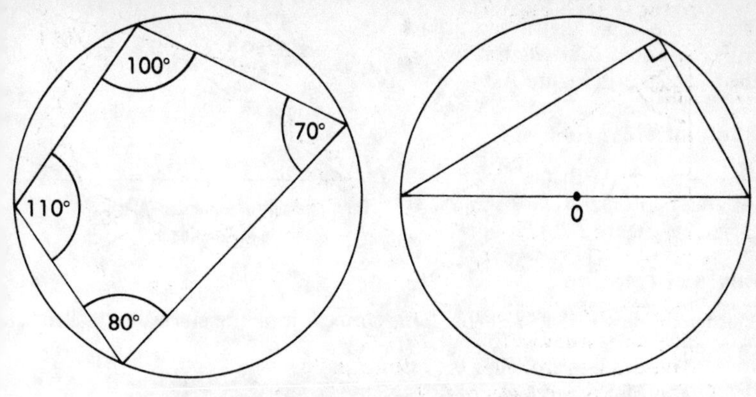

Fig A.11 Fig A.12

12 If you draw any triangle in a semi-circle where one of the sides is the diameter, as in Figure A.12, then the angle at the circumference is a *right angle.*

■ **Exam Question**

Figure A.13 (which is not to scale) represents three sets of parallel lines. State the size of:

a) angle a,
b) angle b,
c) angle c.

(NEA; I)

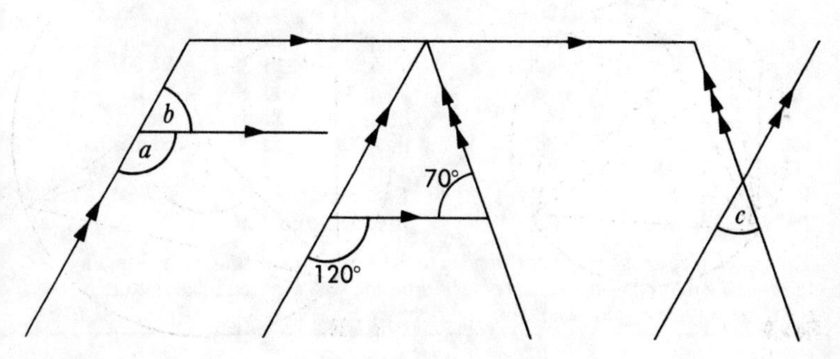

Fig A.13

■ **Solution**

a) $a = 120°$
b) $b = 180 - 120 = 60°$
c) $c = 180 - (70 + 60) = 50°$

Angle of depression

This is the angle that a line makes with a horizontal line *looking down*; this is angle A in Figure A.14.

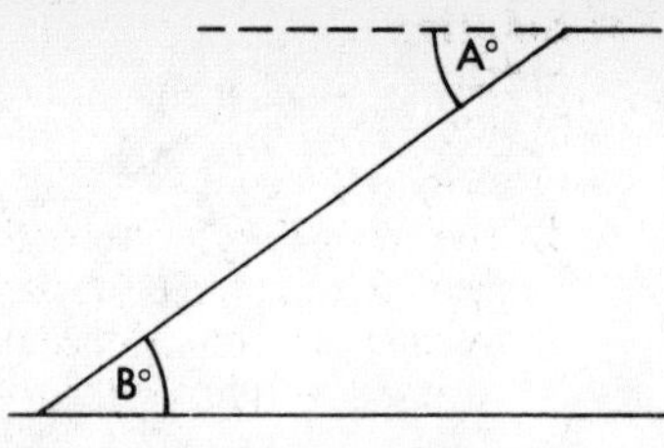

Fig A.14 Angle of depression (A) and angle of elevation (B)

Angle of elevation

This is the angle that a line makes with the horizontal *looking up*; this is angle B in Figure A.14.

Angle of rotation

This is the angle that a shape *turns through* in a transformation called a rotation.

APPLICATION

This is where mathematics is used in 'real life'. An example of application of mathematics is where a real situation is used to give a setting for an examination question.

- **Exam Question**
  Figure A.15 is taken from a Post Office Information Leaflet entitled *Postal Rates*. It shows the cost of sending letters by post.

LETTER POST

Weight not over	*1st class*	*2nd class*	*Weight not over*	*1st class*	*2nd class*
60g	19p	13p	500g	92p	70p
100g	26p	20p	600g	£1.15	85p
150g	32p	24p	700g	£1.35	£1.00
200g	40p	30p	750g	£1.45	£1.05
250g	48p	37p	800g	£1.55	Not
300g	56p	43p	900g	£1.70	admissible
350g	64p	49p	1000g	£1.85	over 750g
400g	72p	55p	Each extra 250g		
450g	82p	62p	or part thereof 45p		

The Post Office aims to deliver (Monday to Saturday) 90% of first class letters by the working day following the day of collection and 96% of second class letters by the third working day following the day of collection.

Fig A.15

a) What is the cost of sending a letter weighing 150g by second class post?
b) What is the cost of sending a letter weighing 180g by first class post?
c) Jayne sends a letter by first class post and it costs her 64p. What is the maximum weight her letter could have been?
d) Desmond wishes to send a packet by letter post. It weighs 1.4kg. What will it cost to send the packet?

(NEA; I)

■ **Solution**

a) 24p
b) 40p
c) 350g
d) The 1000g (1 kg) will cost £1.85. The table then indicates that the next 250g costs 45p (no parts allowed) again and again until the weight is covered. Hence the 400g will need 2 payments of 45p to cover the weight. So the total cost is £1.85 + 90p = £2.75.

APPROXIMATE

The 'intelligent' guess at an answer to some mathematical problem. It is often found by **rounding** the data off, then by working out a simple problem. For example:

> A family went from Bude in Cornwall to York in Yorkshire, a distance of 473 miles, in 7 hours 15 minutes. What was the approximate speed?
>
> Round 473 off to 490 and the time to 7 hours to give a speed of 490/7 = 70 mph.

In a question that specifically asks you for an *approximate* answer you ought to be able to show *how* you obtained your answer by either rounding off and doing the problem, or doing the problem and then rounding off. Of course an approximate answer should not be given to too many significant figures.

ARC

An arc is the curved part of a sector, as in Figure A.16, where O is the centre of a circle, and OA and OB are both radiuses (radii) of the circle. The *arc length* is calculated with the formula:

■ arc length $= \dfrac{X\pi D}{360}$

where D is the diameter of the circle, and X is the angle of the sector.
Questions that use this formula will appear on the Intermediate or Higher level papers only.

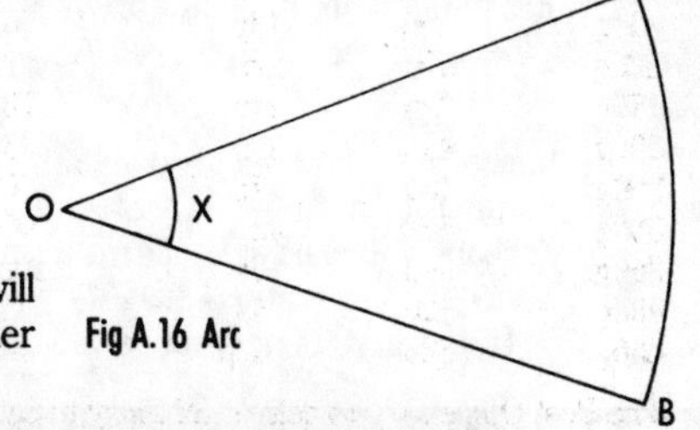

Fig A.16 Arc

AREA

Area is the amount of space inside a flat two-dimensional (2D) shape, and is measured in squares, e.g. square centimetres or square metres.

Common areas

- Area of **rectangle** = length × breadth
- Area of **triangle** = ½ base length × height
- Area of **parallelogram** = base length × height

- Area of **trapezium** = average length of parallel sides × perpendicular distance between them
- Area of **circle** = πr^2, where r is the radius
- Area of **sector** = $\frac{x\pi r^2}{360}$ where x is the angle of the sector.

There will be quite a few examination questions on this topic every year and at every level. These formulae are usually given to you on a *formula sheet* so do ensure that you are familiar with the formula given by your examination group.

- **Exam Question**

  Figure A.17 shows a logo which consists of an equilateral triangle, whose side is of length 4 cm, and three circular arcs, whose centres are the vertices of the triangle.

  a) Find the total area of the region bounded by the three arcs, giving your answer correct to three significant figures.

  b) If the length of each side of the equilateral triangle is $4x$ cm, find an expression for the area in terms of x.

  (NEA; H)

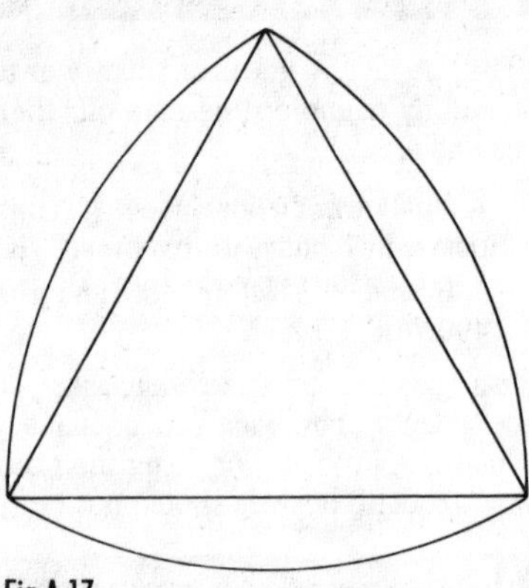

Fig A.17

- **Solution**

  a) The area of the triangle is found by (½ × base × height) (½ × 4 × 4 × sin 60° = 6.9282).
  Each segment = sector area − triangle area
  Where the sector area = $\frac{60}{360} \times \pi \times 4 \times 4 = 8.37758$

  hence segment area = 1.44938
  So the total area = 3 × segment + triangle = 11.3 cm^2.

  b) A shape with said side of $4x$ is a similar shape with length ratio of $1:x$, hence the area ratio is $1:x^2$.
  Hence area of new shape will be $11.3 \times x^2 = 11.3x^2$.

ARITHMETIC MEAN

◀ Mean ▶

ASYMPTOTE/ASYMPTOTIC

Usually refers to a graph where the curve gets nearer and nearer to a straight line (usually one axis) but never quite touches that line. A **hyperbola**, for example, has two asymptotes.

◀ Graphs, Reciprocal equations ▶

AURAL TESTS

An aural test is one in which you have to listen to the questions and then answer them. You do not have a printed paper with the questions on. You need mental arithmetic or a quick pen and pencil routine, since you will not be allowed to use your calculator. The question will usually be read out twice and slowly so that you can jot down some information. Jot down the necessary *figures*, but as few *words* as possible, since you do need to be *listening* to the whole question. For example:

Each question will be read out twice, slowly. Then you will be given one minute in which to arrive at your answer and to write it down.

1 A rectangle measures 9 cm by 4 cm. A square has the same area. What is the perimeter of this square?
2 What is the cost of 30 square metres of lino at £4.50 per square metre?
3 A salesman earns 2% commission on sales of £3000. What is his commission in pounds?
4 The average cost of 6 drinks is 92p. What is the total cost?
5 If one unit of gas cost 27p, what is the cost in pounds of 300 units?
6 A revise guide is 1.5 cm thick. How many of these books can be placed on the shelves of a bookshop if the shelves are one metre long?
7 Which of the following is the nearest to the average height of a woman in Europe: 98 cm; 160 cm; 198 cm?

AVERAGE

What people usually mean by the average is the 'middle thing' of the thing that most people have. There are three different types of average:

Mode

This is what most people have or do, or the most frequently appearing item of data. For example, the record that is currently number 1 in the charts is the one that has sold the most during the previous week, so it is the *mode* or the modal record.

Median

This is the 'middle' item of data once it has all been put into a specific order. For example, if you had 9 people and wanted to find their *median* height, then you would put them into the order of their heights; whoever is in the middle then has the median height. If you need to find the median of an *even* set of numbers, then you need the arithmetic mean of the *middle two* numbers.

Mean (arithmetic mean)

This is the average that most people are familiar with and really intend when they use the word 'average'. It is found by adding up all the data and dividing it by the number of data items you added up. For example:

From the numbers 1,2,3,4,4,5,5,5,5
the **mode** will be 5 (the number occuring most often)

the **median** will be 4 (the middle number, with numbers arranged in order)
the **mean** will be 3.78 (34/9, i.e. the total divided by the 9 items of data)

The most common mistake to make with averages is to calculate the wrong type; so do learn the difference between each type of average.

■ **Exam Question**
Two dice were thrown together 125 times. The scores obtained are shown in Figure A.18. Use this information to find:

a) the mode of the scores,
b) the median score,
c) the mean score.

(NEA; I)

Score	*Frequency*
2	4
3	7
4	9
5	11
6	7
7	5
8	17
9	19
10	24
11	12
12	10

Fig A.18

■ **Solution**
a) The mode is the most common, hence the 10.

b) The median is the $\frac{125 + 1}{2} = 63$rd number, so counting down to the 63rd number you come to 9.

c) The mean score is calculated as:
$(2 \times 4) + (3 \times 7) + (4 \times 9) + (5 \times 11) + (6 \times 7) + (7 \times 5) + (8 \times 17) + (9 \times 19) + (10 \times 24) + (11 \times 12) + (12 \times 10)$ all divided by 125. This is
$996 \div 125 = 8.0$
(rounded off).

AXES

The lines on which we put numbers on a graph are called the axes. Usually the *horizontal line* is called the *x*-axis, and the *vertical line* is called the *y*-axis.

AXIS OF SYMMETRY

An axis of symmetry is the line that a 3-dimensional shape rotates around to demonstrate **rotational symmetry**. An example is shown in Figure A.19.

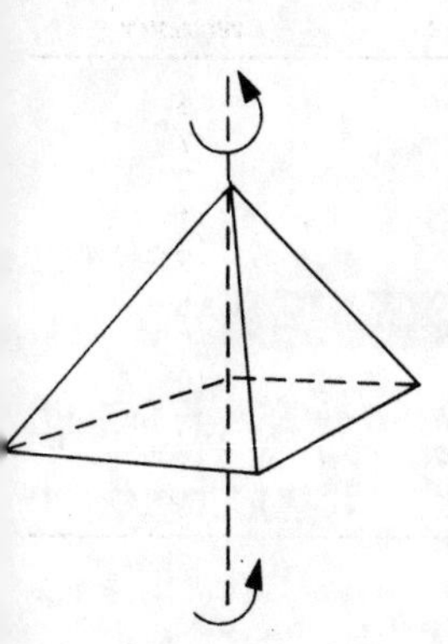

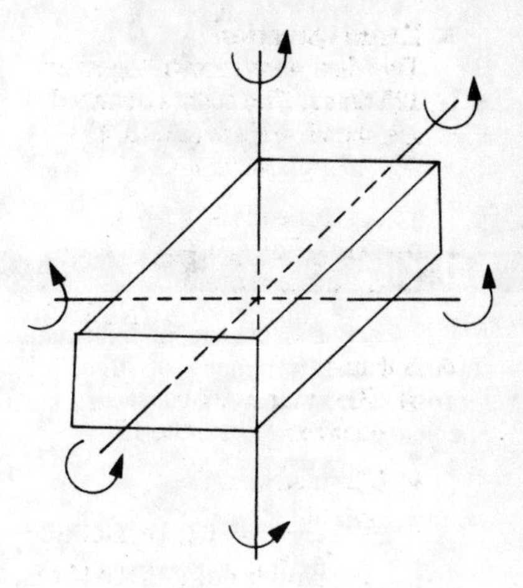

Fig A.19 Axis of symmetry

BAR CHART

A bar chart is a display of information using bars of different *lengths* to represent the frequency of items of data. The bar chart in Figure B.1 represents how many babies were born in one week at Jessop's hospital in Sheffield.

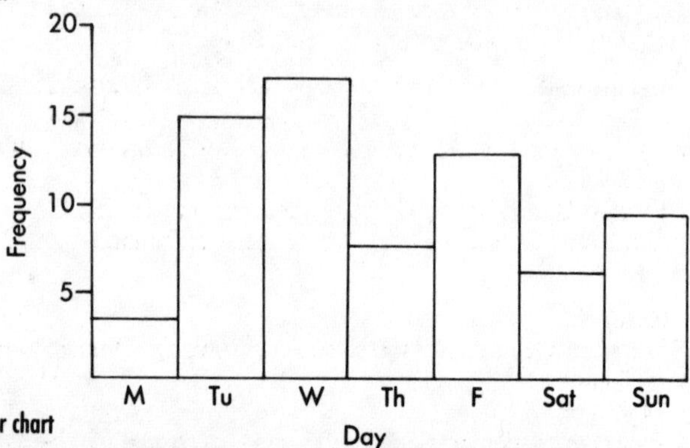

Fig B.1 Bar chart

BASE NUMBER

The number which is raised to some index or exponent.

◀ Index ▶

BEARINGS

Look at a compass similar to the one shown in Figure B.2; you may well see numbers like 005°, 070°, 340°, etc. These are 3-figure *bearings* and represent the *clockwise angle from North* which that direction is making. You should see that east is 090°, south is 180°, south west is 225° etc. The most common error to make in bearings is to measure anti-clockwise instead of clockwise.

Back bearings

If the bearing from A to B is known, then the bearing back from B to A can be

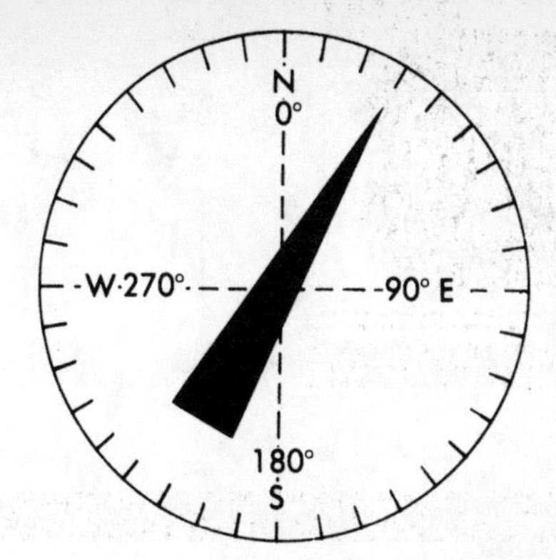

Fig B.2 Bearings

calculated by adding on 180°. If the back bearing is calculated as greater than 360, then subtract 360 from the final answer for the correct bearing.

- **Exam Question**

  A ship sails from point A on a bearing of 035° to point B, 50 km away. From B it then sails 80 km due south to point C.

  a) Using a scale of 1 cm to represent 10 km, make a scale drawing of the ship's course.
  b) Use your scale drawing to find the distance, in km, from A to C in a straight line.
  c) From your scale drawing find the bearing of C from A.
  d) *Calculate* the distance from A to C in a straight line.

  (NEA; I)

- **Solution**

  a) You should have drawn the ship's journey using ruler and protractor.
  b) 48.5 km.
  c) 144°.
  d) By a combination of trigonometry and Pythagoras, on the right-angled triangles formed with Point D the perpendicular of A to DC.

Then;
$$BD = 50 \sin 55° = 40.96$$
$$AD = 50 \cos 55° = 28.68$$
$$DC = 80 - BD = 39.04$$
$$AC^2 = AD^2 + DC^2 = 2346.8$$
$$AC = \sqrt{(2346.8)} = 48.44 \text{ km}$$

BEST BUY

You face a best buy problem when there are alternatives and you have the choice of which to buy. The usual technique is to find either the ***weight per penny*** (or £1) of each item or the ***cost per unit weight*** (or length, or area, etc.). This choice will usually depend on the information presented. The best buy will then be either the one that gives you the most weight etc. for each unit of money, or cost least per unit of weight etc.

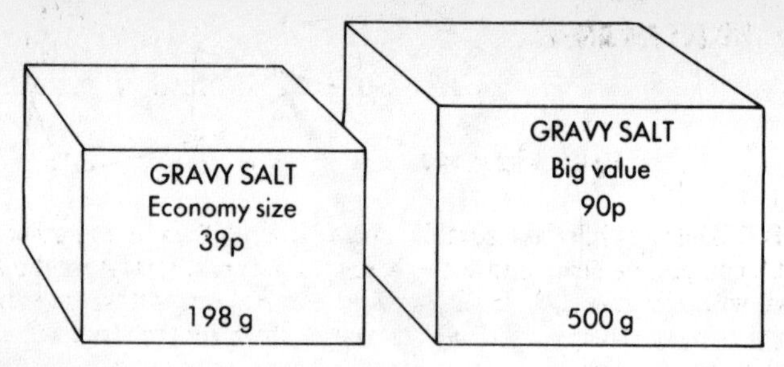

Fig B.3 Best Buy

■ **Exam Question**
Gravy salt is sold in two different sized packets as shown in Figure B.3.

a) What is the weight of the gravy salt per penny
  i) for the 'Economy' packet?
  ii) for the 'Big value' packet?
b) From your answer to part a), which packet is the better value, and why?

(NEA; I)

■ **Solution**
a) i) 198 g ÷ 39 = 5.0769... g
  ii) 500g ÷ 90 = 5.555... g
b) The 'Big value' is the better buy because you get more weight for your money.

BISECT

Bisect means to put into two equal halves. It is most commonly used in **constructions** as one of the following:

BISECT A LINE

Figure B.4 illustrates how to bisect the line AB. Set your pair of compasses to about three quarters of the length of line AB, and with the sharp point at one end draw a faint semi-circle; then repeat from the other end, keeping the compass arc the same length. You need to find out where the two semi-circles cross over. The straight line between these points will give the *line bisector*.

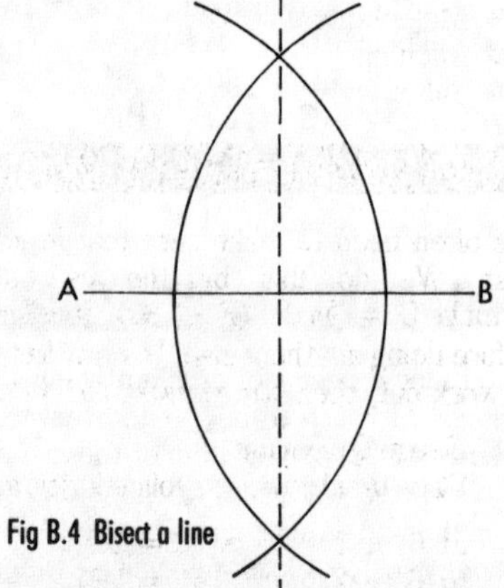

Fig B.4 Bisect a line

BISECT AN ANGLE

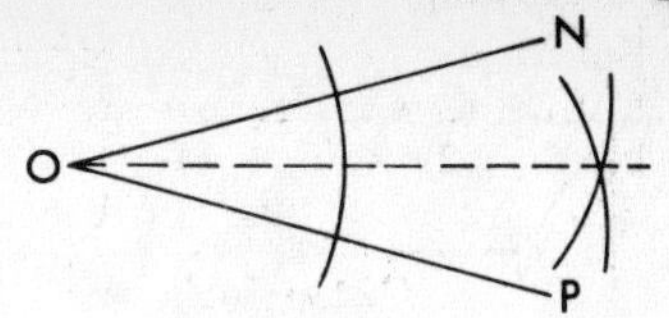

Fig B.5 Bisect an angle

Figure B.5 illustrates how to bisect the angle NOP. With your compasses set at about 2 cm, put the sharp end at the vertex of the angle O, and arc the angle as shown with a faint line. At both points where this arc cuts the *sides* of the angle, and using the sharp end of the compasses, draw another faint arc across the angle as shown. Where these last two arcs *cross*, join to the vertex at O. The resulting line is the *angle bisector*.

An examination question will use the words 'with compass and pencil only, construct...'. You then must construct the bisectors as shown, or gain no marks at all. By all means check your accuracy afterwards with a protractor or ruler and redraw if you are not accurate. Also do remember to show your construction lines clearly and not to rub them out or to draw them so faint that no one can see them.

BODMAS

This is a mnemonic which helps us remember what to work out first in an arithmetical situation where there seems to be several ways of proceeding. It stands for the phrase:

Brackets, Of, Division, Multiplication, Addition, Subtraction.

We do these things in that order. For example:

$3 \times 4 + \frac{1}{2} \text{ of } 12 \div 3 - (9 - 7)$

is done in the order:

■ Brackets	3 × 4	+ ½ of 12	÷ 3	− 2
■ Of	3 × 4	+ 6	÷ 3	− 2
■ Division	3 × 4	+ 2		− 2
■ Multiplication	12	+ 2		− 2
■ Addition		14		− 2
■ Subtraction		12		

BRACKETS

We often need to make sure that in formula certain numbers are calculated first. We do this by the use of brackets. For example in the formula $C = \frac{5}{9} \times (F - 32)$, it is important to subtract the 32 from F before doing anything else. If a bracket appears in a formula or an expression to work out, then *always work out the bracket first.*

- **Exam Question**

  Place brackets in the following statements to make them true.

  a) $6 \div 3 + 5 \times 4 = 3$
  b) $6 - 2 \times 5 + 4 = 36$

(MEG; H)

■ **Solution**

a) $6 \div (3 + 5) \times 4 = 3$

b) $(6 - 2) \times (5 + 4) = 36$

CALCULATOR

All GCSE examinations allow you to have your calculator with you for all the written papers and for some of the coursework assignments. The questions will be set on the assumption that you have got a suitable calculator with you. The calculator is the responsibility of the *candidate* and not the school, so candidates need to supply their own, together with spare batteries if required. Always try to use a *familiar* calculator, since there may be considerable differences between models. You do not want the extra pressure of learning to use a calculator in a timed situation.

When using a calculator in the examination, do not forget to write down any *method of solution* that is relevant. Otherwise marks could be lost if the answer is wrong and no method of solution can be seen. Try also to keep as accurate an answer in the calculator as you can for *multiple-stage* calculations; any rounding off should be done at the final stage. If you round off too early, your final answer will be less accurate.

CANCELLING

This is what we call the process when a fraction is divided top and bottom by a particular **integer** (whole number). For example, $9/15$ cancels to $3/5$ by dividing both top and bottom by 3.

CARTESIAN GRID

◀ Graph ▶

CENTRE OF ENLARGEMENT

◀ Enlargement ▶

CHARTS

An important part of the GCSE examinations is the inclusion of charts that are a familiar sight in everyday life.

Conversion chart

Figure C.1 shows a conversion chart. This particular chart converts from litres to gallons, and of course the other way round.

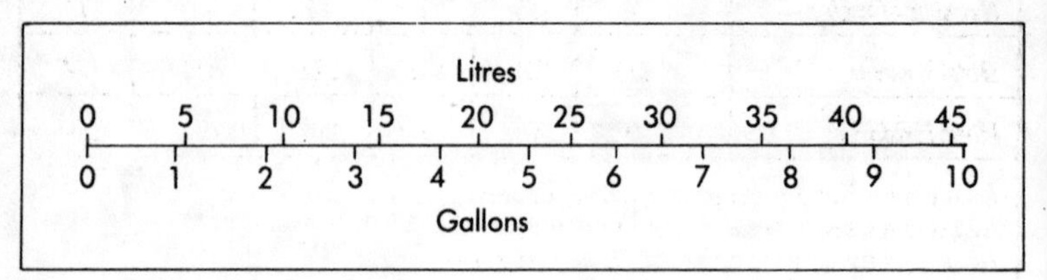

Fig C.1 Litres to gallons conversion chart

Information chart

The type of chart in Figure C.2 needs to be read carefully to provide information. You would read this particular chart to find out the details about certain monsters in the game of Dungeons and Dragons.

Monster	*Size*	*To Hit A.C. 0*	*Armor class*	*Hit dice*	*No. of attacks*	*Damage per attack*
diplodocus	L	7	6	24	1	3-18
elasmosauras	L	8	7	15	1	4-24
gargosaurus	L	9	5	13	3	1-3/1-3/7-28
iguanadon	L	13	4	6	3	1-3/1-3/2-8
lambeosaurus	L	9	6	12	1	2-12
megalosaurus	L	9	5	12	1	3-18
monoclonius	L	12	3/4	8	1	2-16
mosasaurus	L	9	7	12	1	4-32
paleascincus	L	12	−3	9	1	2-12
pentaceratops	L	9	2/6	12	3	1-6/1-10/1-10
plateosaurus	L	12	5	8	nil	nil
plesiosaurus	L	7	7	20	1	5-20
pteranodon	L	16	7	3+3	1	2-8
stegosaurus	L	7	2/5	18	1	5-20
styracosaurus	L	10	2/4	10	1	2-16
teratosaurus	L	10	5	10	3	1-3/1-3/3-18
triceratops	L	7	2/6	16	3	1-8/1-12/1-12
tyrannosaurus rex	L	7	5	18	3	1-6/1-6/5-40

Fig C.2 Dungeons and Dragons monster chart

The chart in Figure C.3 is another information chart, telling you the various costs of staying at a hotel for each week during the year.

Questions involving charts mainly appear at Basic and Intermediate level but could also appear on the Higher level papers.

◀ Pie chart ▶

Basic holiday price in £s per person – Gatwick departures						
Between	9 May 12 June		13 June 24 July		25 July 5 September	
Number of nights	7	14	7	14	7	14
Hotel Knowle	211	319	219	328	224	367
Hotel Hildaro	193	276	197	285	204	294
Addition for Heathrow departure £30 per person Addition for sea views £1.90 per person per night Addition for insurance cover £4.75 per person No reduction for children or OAPs						

Fig C.3 Holiday cost chart

CHORD

Any straight line drawn in a circle from one part of the circumference to another is called a chord. As you can see in Figure C.4, a chord divides a circle in two segments, the larger one called the *Major segment* and the smaller one the *Minor segment.*

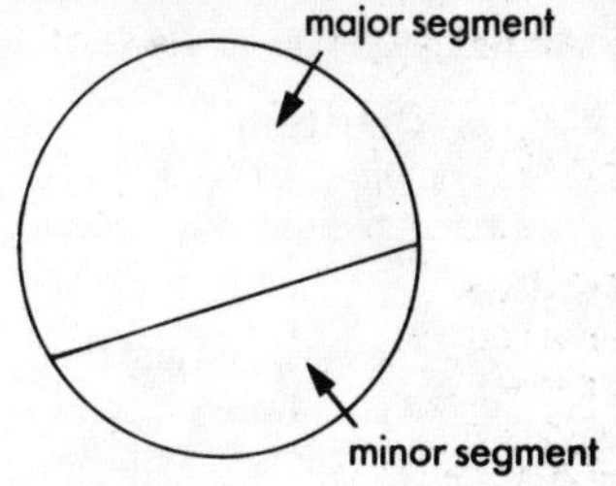

Fig C.4 Chord

Angles on a chord

◀ Angles ▶

Intersecting chord theorem

In the ciricle in Figure C.5 where two chords AB and CD intersect at X, then AX.BX = CX.DX. This will even be true where the chords intersect *outside* the circle.

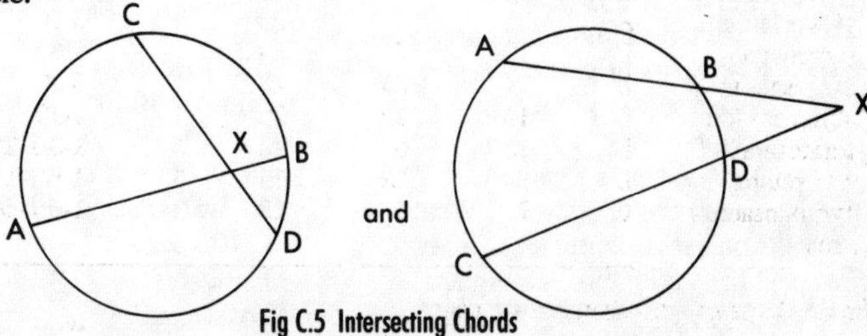

Fig C.5 Intersecting Chords

CIRCLES

A circle is the locus of a point that is always the same distance from a given point which is the centre of that circle. You should learn the terms associated with circles (Fig C.6).

- The outside edge of the circle is called the **circumference.**
- Any line from the centre of the circle to the circumference is called a **radius.**
- Any line through the centre of a circle from circumference to circumference is called a **diameter.**
- The *length of the circumference* of a circle (**perimeter**) is given by the formula: circumference = $\pi \times$ diameter.

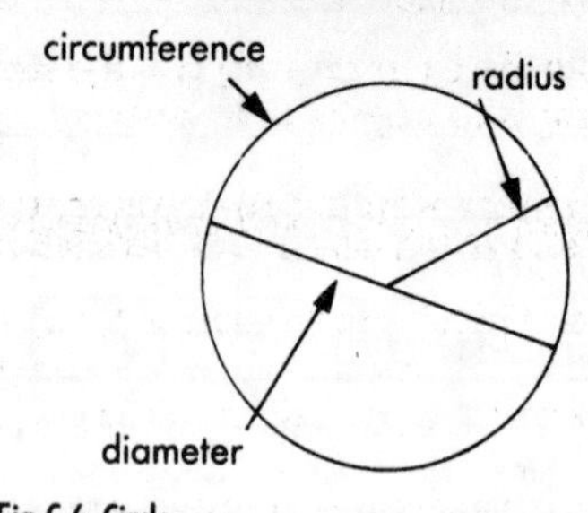

Fig C.6 Circle

- The *area* of a circle is given by the formula: area $= \pi \times (\text{radius})^2$.

Angles in a circle

◀ Angles, Pi (π) ▶

- **Exam Question**
  In a computer game, a ghost eats circles and triangles. Each time it eats a circle its brightness is increased by 5 watts per cm^2 of circle area eaten (Fig C.7).

  a) A game is played and the ghost eats twelve circles each of radius 0.5 cm.
  i) What is the area of each circle?
  ii) By how much has the brightness of the ghost increased after eating all the twelve circles?

  b) Each time a ghost eats a triangle its brightness is decreased by $\frac{2}{5}$ of its brightness.
  A ghost starts a game with a brightness of 50 watts. What would its brightness be after it had eaten three triangles?
  (NEA; I)

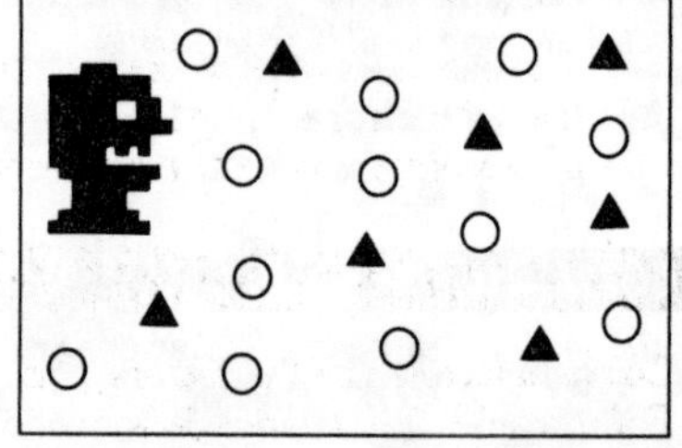

Fig C.7 'Ghost eater'.

- **Solution**
  a) i) $\pi \times (0.5)^2 = 0.78\ cm^2$ (or 0.79)
  ii) $0.78 \times 12 \times 5 = 47.1$ watts (using unrounded 0.78...)
  b) $50 \times \frac{3}{5} \times \frac{3}{5} \times \frac{3}{5} = 10.8$ watts.

CIRCUMFERENCE

The circumference of a circle is the outside edge and is calculated by the formula:

- circumference $= \pi \times$ diameter

The circumference is often also quoted with the formula:

- circumference = 2 × π × radius

Both these formulae will give the same length of circumference and a common error is to mix the two together.

COEFFICIENT

A coefficient is the number in front of a letter or group of letters in an algebraic term. For example in $3x$, the 3 is the coefficient of x. In $5xy$ the 5 is the coefficient of xy. The coefficient is multiplied by the letters.

COMBINED EVENTS

A combined event is part of **probability** where two or more events are happening at the same time. The two main situations which are examined at GCSE levels are concerned with the AND rule and the OR rule:

- **AND**
  AND is the type where both events happen at the same time. To find this combined probability we *multiply* the probabilities of each single event.
- **OR**
  OR is the type when either one event or the other can happen, but not both at the same time. To find this probability we *add* the probabilities of each event.

  The events must all be part of a *sample space*, that is to say a situation that takes account of **ALL** possible outcomes. This is often illustrated on a **Tree diagram** or **Venn diagram**.

COMMON FACTORS

Common factors are the factors that two or more **integers** have in common. For example, the common factors of 18, 24 and 36 are 1, 2, 3, and 6. Here, 6 is the **highest common factor** (HCF).

COMMON MULTIPLES

Common multiples are the multiples that two or more **integers** have in common. For example, the common multiples of 5 and 3 are 15, 30, 45... Here, 15 is the **lowest common multiple** (LCM).

COMPLEMENT

The complement of a set A is everything in the given **Universal set** that is *not* in the set A. The notation used for the complement of set A, is A′. For example:

Where $\mathscr{E} = \{1, 2, 3, 4, 5, 6, 7, 8, 9\}$ and $A = \{3, 4, 5, 6, 7\}$
then $A' = \{1, 2, 8, 9\}$

COMPOSITE FUNCTION

◀ Function ▶

COMPOUND INTEREST

This is the type of interest most likely to be paid by banks and building societies. It is a system that allows your interest to grow, giving interest on the interest. For example:

To find the compound interest on £150 at a rate of 8% per annum (that is per year) for 3 years.

At the end of 1 year, interest = $150 \times 8/100$ = £12
So total amount after 1 year = £150 + £12 = £162
End of 2nd year interest = $162 \times 8/100$ = £12.96
So total after 2 years = £162 + £12.96 = £174.96
End of 3rd year interest = $174.96 \times 8/100$ = £14.00 (rounded to nearest penny).

The most efficient way to calculate compound interest is:

where R is the interest rate,
where N is the number of times the rate is to be applied,
where P is the amount started off with,

then the *final amount* is given by $P \times (1 + R/100)^N$.
To find the *interest paid*, simply subtract the original amount from the final amount.

- **Exam Question**
  Elsie won £60 in a beauty contest and put it into a building society account that paid 8% compound interest annually. How much would she have in this account if the money was left there for 3 years?

- **Solution**
  R = 8
  N = 3
  P = £60
  Final amount $= P \times (1 + R/100)^N$
  $= 60 \times (1 + 8/100)^3$
  = £75.58

We can of course explain the process involved in rather more detail. After 1 year the **simple interest** will be $(60 \times 8 \times 1) \div 100$ = £4.80, which is added to the account. So the second year starts with a total amount of £64.80. At the end of the second year the simple interest will be (£64.80 × 8 × 1) ÷ 100 = £5.184, which would be rounded off to £5.18 and added to the £64.80 to give a new total amount of £69.98. At the end of the third year the simple interest will be (£69.98 × 8 × 1) ÷ 100 = £5.5984, which would be rounded off to £5.60 which, when added to £69.98, will give a final figure of £75.58.

■ **Exam Question**
During the first few weeks of its life, the octopus in Figure C.8 increases its body weight by 5% each day. The octopus was born with a body weight of 150 grams. How much will it weigh after:

a) 1 day?
b) 3 days?

(NEA;I)

Fig C.8

■ **Solution**
a) It will weigh $150 \times 1.05 = 157.5$ g
b) The quickest solution is to calculate $150 \times (1.05)^3$, which is 173.6438g. A sensible rounding off would be 174g.

CONE

A cone is a three-dimensional shape with a circle at the base and a smooth curved surface rising to a point at the top, like a witch's hat (Fig. C.9).

Where the cone of height h has a base radius r and a slant height l, then:

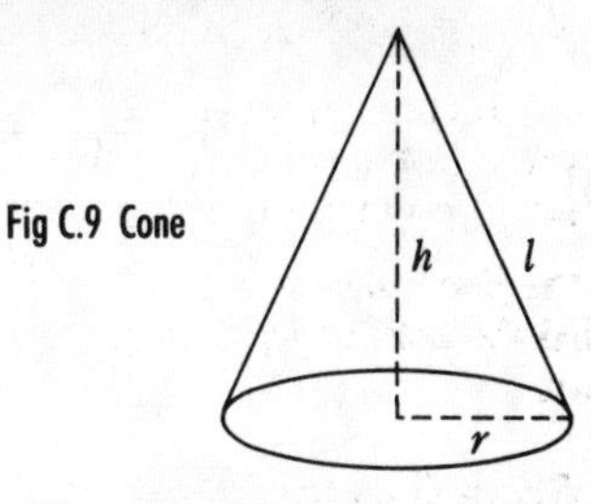

Fig C.9 Cone

- the *curved surface area* (CSA) is given by the formula CSA $= \pi rl$
- the *volume* is given by the formula $V = \frac{1}{3}\pi r^2 h$

CONGRUENCE/CONGRUENT

Two shapes are congruent if they are exactly the same shape and size, so that one shape would fit exactly on top of the other. This means that the angles of one shape would be the same as the other shape. It also means that all the *lengths* will be the same in one shape as in the other.

■ **Exam Question**
State which **two** shapes in the Figure C.10 are:

a) congruent,
b) similar but not congruent.

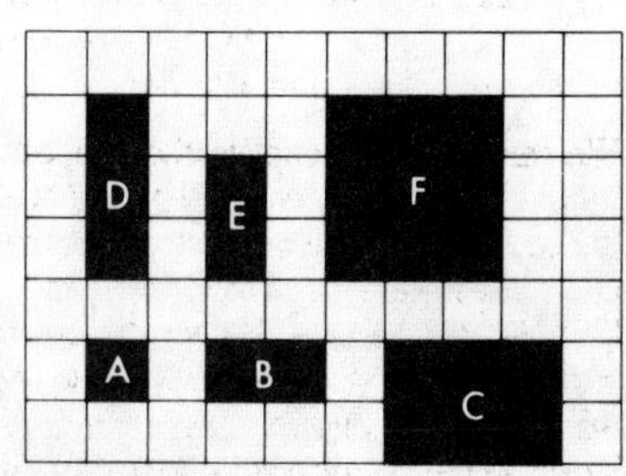

Fig C.10 Congruent shapes

■ **Solution**
a) shapes B and E
b) shapes A and F

CONGRUENT TRIANGLES

Within the GCSE examination, candidates need to know when two *triangles* are congruent or not. The following examples illustrate the minimum information needed to determine whether two triangles are congruent or not:

- In Figure C.11, all three sides are the same in each triangle, so ABC will be congruent to XYZ.
- In Figure C.12, all three angles and one corresponding side are equal, so DEF will be congruent to PQR.
- In Figure C.13, two sides and the included angle are the same, so JKL will be congruent to STU.

◀ Similarity ▶

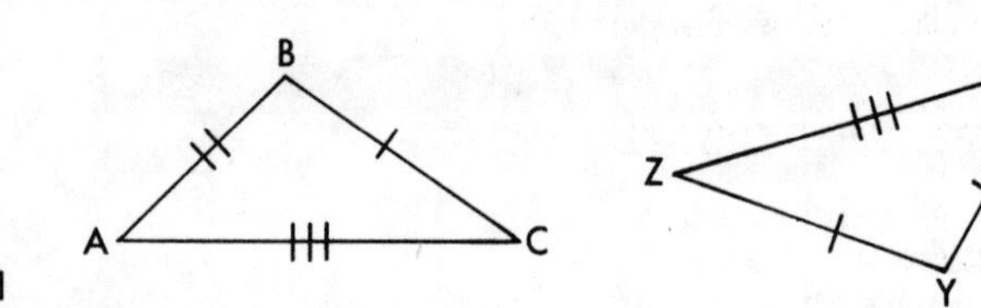

Fig C.11

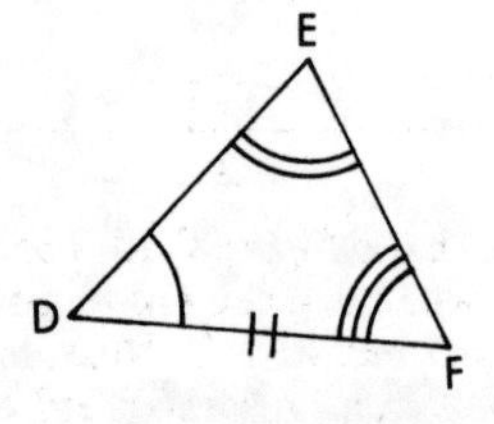

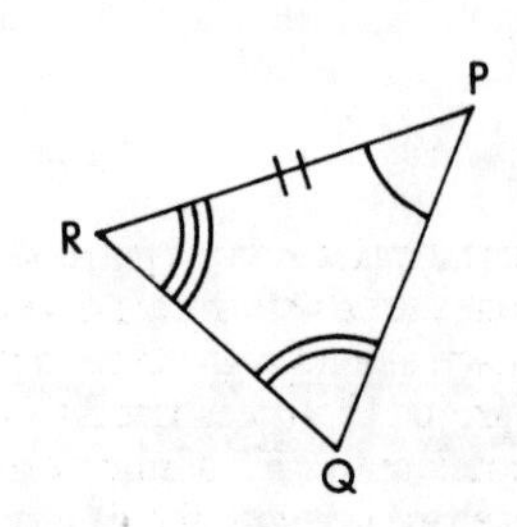

Fig C.12

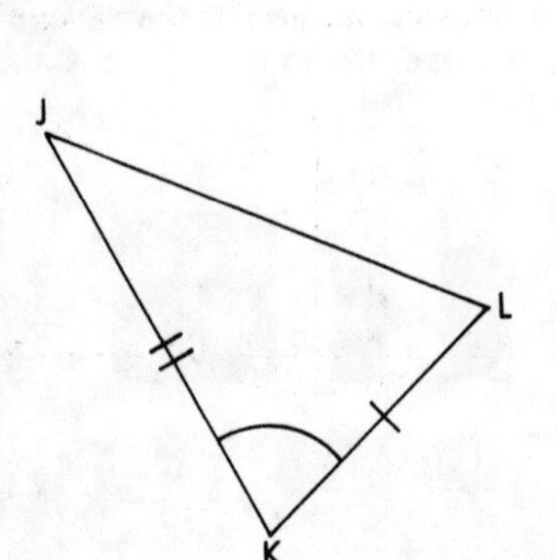

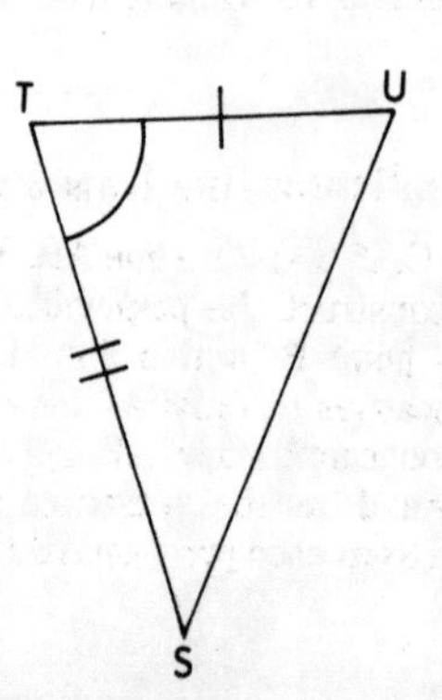

Fig C.13

CONSTANTS

Constants are things that do *not* change. For instance in $x = 6y - 3$, the 3 is always 3 while the 6y will change as y changes. So here the 3 is called the *constant* term.

CONSTRUCTIONS

Constructions are where you need to use a pair of compasses, a pencil and a straight edge, which could be a ruler if you need to construct precise sizes.

Notice that you are generally asked to construct *without* a protractor. However you could be asked to construct a shape containing an awkward angle, such as 51^0, and then you *would* use one.

The constructions asked within the GCSE will be:

- Line bisectors. ◀ Bisectors ▶
- Angle bisectors. ◀ Bisectors ▶

SPECIALISED CONSTRUCTIONS

A right angle

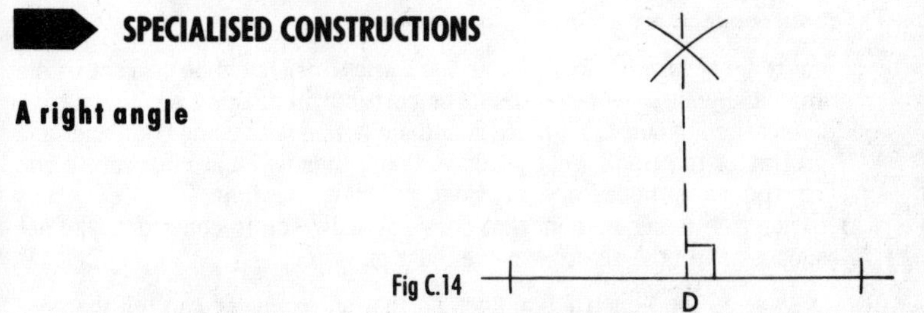

Fig C.14

To construct a *right angle* at the point D on the line in Figure C.14, use a pair of compasses with an arc of about 4cm. Then, with the sharp end on point D, arc each side of point D to give two marks *equidistant* from D. Next, extend the arc on the compasses to about 6 or 7cm. Putting the sharp end of the compass onto *each of these two marks in turn*, draw an arc *over* the point D. These two last arcs should cross over each other, as in the diagram. Now join up the point D to where these arcs have crossed and you have your right angle.

A perpendicular line from a point

In Figure C.15 there is a line AB. We wish to construct the *perpendicular line* from point P to this line. Use your compasses to draw as wide an arc as you can cutting AB at two points, with P as the centre of the arc. Then from each point on AB that

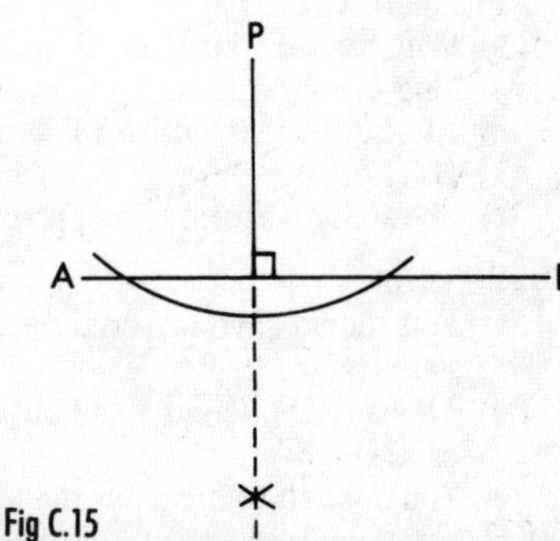

Fig C.15

you have arced, and using the *same* compass opening, make an arc *under* the line AB as shown. The two arcs will cross, giving you the point from which to draw the perpendicular line from P to the line AB.

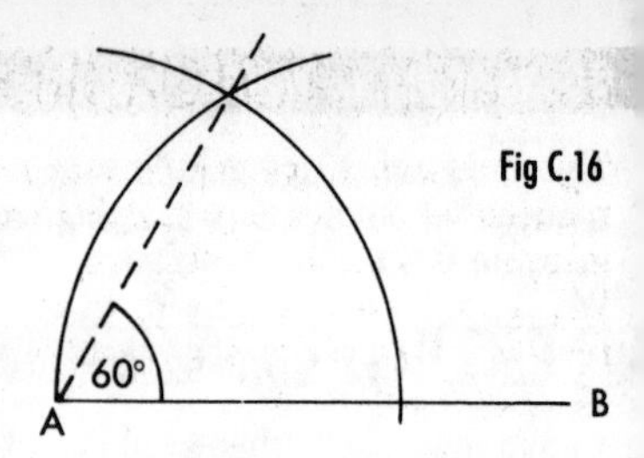

Fig C.16

A 60° angle

To construct a 60° angle at point A on the line AB shown (Fig C.16), you can set your compass to any distance you like. Then with the sharp end at point A, draw faintly the quarter circle that arcs through AB. Where this has cut AB, put the sharp end of the compasses (keep it the *same* distance again) and draw the arc faintly that goes from A and cuts through the first arc. Where these two arcs *cross* you can draw a straight line to point A and you have your angle of 60°.

The most common errors in constructions are:

a) not to be accurate enough. Your *measurements* should be correct within 1mm and your *angles* also should be correct within 1 degree. If you have been asked to construct there is nothing wrong with using your ruler and protractor to **check** what you have done; then if you are inaccurate you can redraw as necessary, provided you have the time.
b) to actually use equipment that does not allow you to construct, like set squares and protractors, or even just guessing!

Always *show* all the construction lines so that an examiner can tell that you have constructed. If there are no visible lines of construction, then the examiner will assume that you have used other means and give you no marks at all.

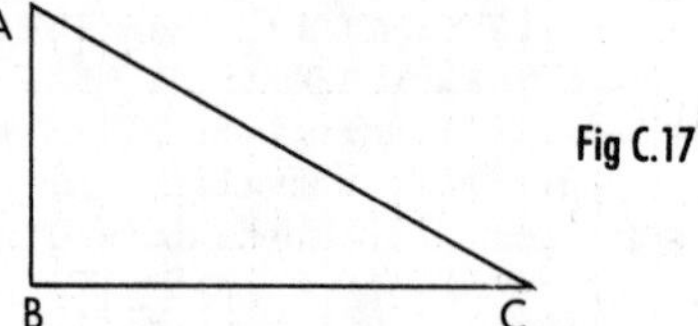

Fig C.17

- **Exam Question**

  In Figure C.17:

  a) Construct the bisector of angle BAC. Make sure the bisector is drawn long enough to meet BC and label this intersection X.
  b) Measure the lengths of BX and XC, and hence write down the ratio BX:XC.
  c) Show that this ratio is approximately the same ratio as AB:AC.

- **Solution**

  a) You should have bisected the angle using a pair of compasses and **not** a protractor.
  b) Measure the lengths and put to a ratio, which should be close to 3:5 or 1:1·6667
  c) You needed to measure the lengths of AB and AC, put this as a ratio and you should show this also to be close to 1:1·6667.

CONTINUOUS DATA

This is data that *cannot* be measured exactly and is therefore given to some rounded off amount e.g. people's height, weight and age. In every case the measure is suitably rounded off.

Where this type of data is used in statistics, then its nature has to be recognised when drawing charts or interpreting them.

◀ Discrete data ▶

CONVERSION GRAPHS

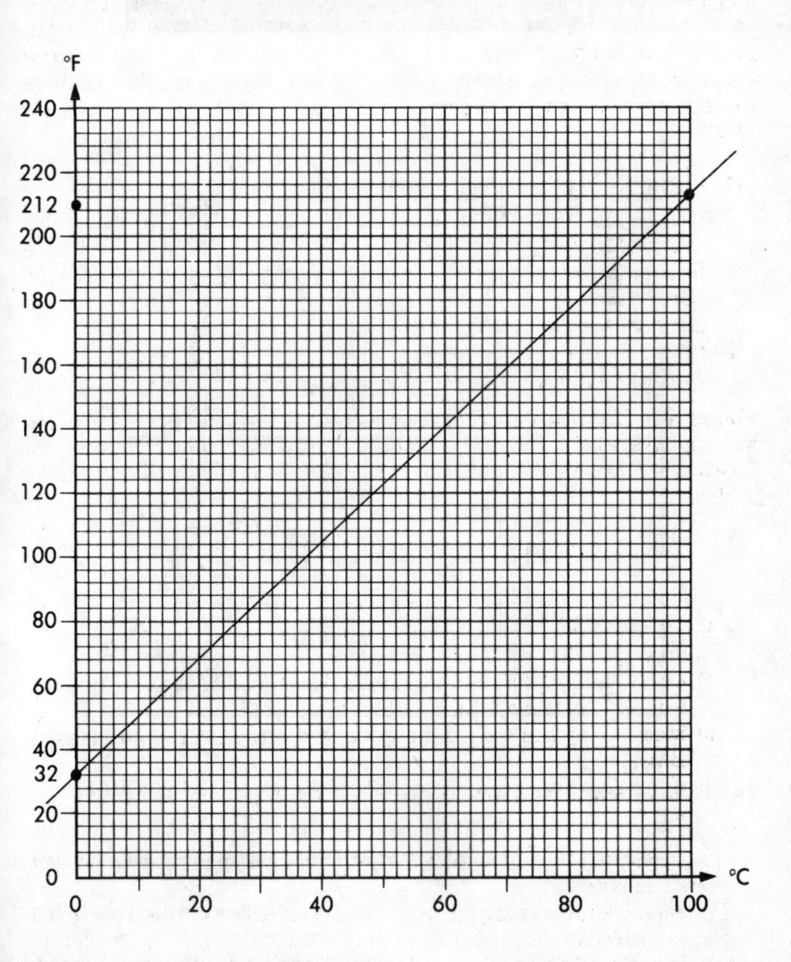

Fig C.18 Conversion graph: Centigrade to Fahrenheit

To help *convert* from one unit to another it is often helpful to have a *graph* which shows the conversion. Figure C.18 shows a Centigrade to Fahrenheit conversion graph. You can find the approximate number of degrees Fahrenheit equal to any number of degrees Centigrade, for example, 20 degrees Centigrade. From the 20 degrees Centigrade mark follow the vertical line up to the graph and then follow it horizontally to approximately 68 degrees Fahrenheit. It can also be worked the other way round to show that 100 degrees Fahrenheit is approximately equal to 38 degrees Centigrade.

There are many different types of *conversion graphs*; for instance, from imperial units to metric units, from one currency to another, and so on.

CO-ORDINATES

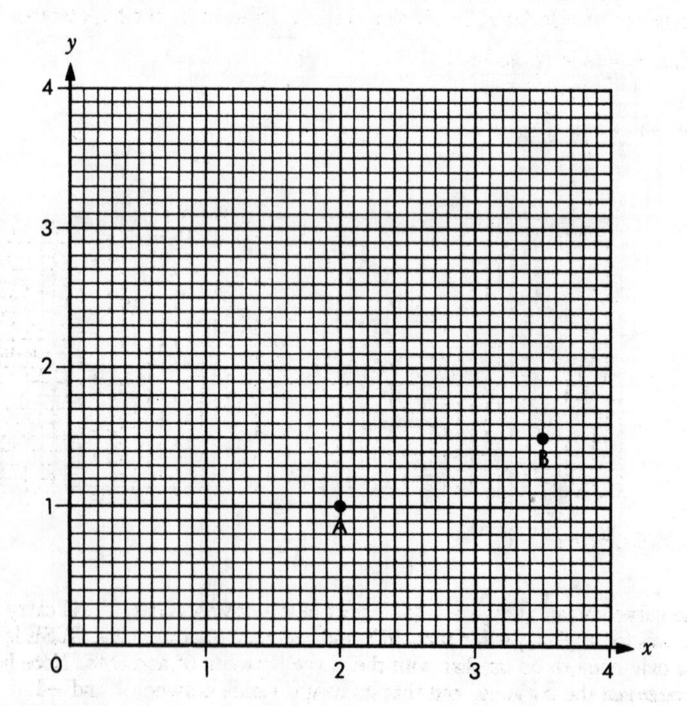

Fig C.19 Co-ordinates

Co-ordinates are pairs of numbers that fix a particular position on a grid with reference to some origin.

Cartesian co-ordinates are the usual ones, where the axes used are at right angles to each other. These axes go through the origin and have the numbers marked on them. The *horizontal* axis is usually called the x-axis, and the *vertical* one the y-axis.

In a co-ordinate we place the number representing the horizontal axis *before* that representing the vertical axis. For example, in Figure C.19:

- The origin in the diagram is the co-ordinate (0,0).
- The co-ordinate of point A is (2,1).
- The co-ordinate of point B is (3.5, 1.5).

The most common mistake made with co-ordinates is to plot them the wrong way round. Another is to plot them more than 1mm away from where they should be plotted.

COSINE

Cosine is part of the topic **trigonometry**, and is usually abbreviated to cos. Every angle has a cosine. When the angle is in a *right-angled triangle*, the cosine can be calculated by dividing the side adjacent by the hypotenuse.

$$\text{Cos } x = \frac{\text{Side Adjacent}}{\text{Hypotenuse}}$$

Cosine curve

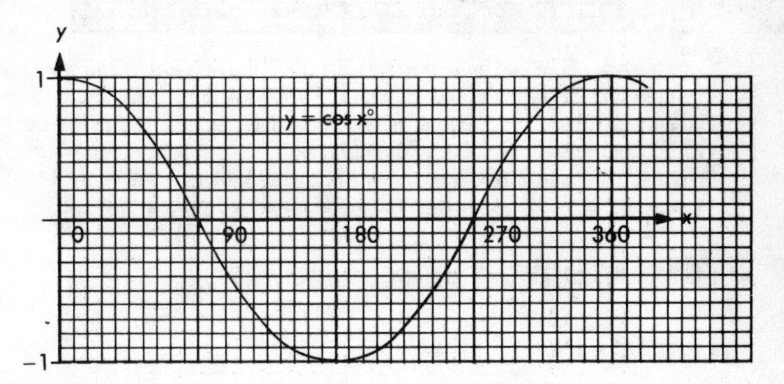

Fig C.20 Cosine curve

The curve shown in Figure C.20 is part of the **cosine curve**. It will carry on like this for angles greater than 360° and less than 0°. But at the GCSE level you only need to be familiar with the curve between 0° and 360°. Note how *symmetrical* the curve is, and that its height varies between 1 and −1.

Cosine rule

This can be used to find a missing length or angle in a triangle (*not* a right-angled triangle):

- If all three sides are known, as in Figure C.21, then any missing angle can be found.

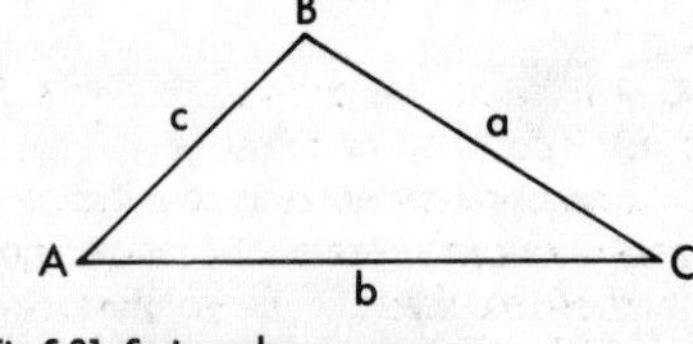

Fig C.21 Cosine rule

Where the three sides a, b and c are known, then use the cosine rule as:

$$\cos A = \frac{b^2 + c^2 - a^2}{2bc}$$

$$\text{or} \quad \cos B = \frac{a^2 + c^2 - b^2}{2ac}$$

$$\text{or} \quad \cos C = \frac{a^2 + b^2 - c^2}{2ab}$$

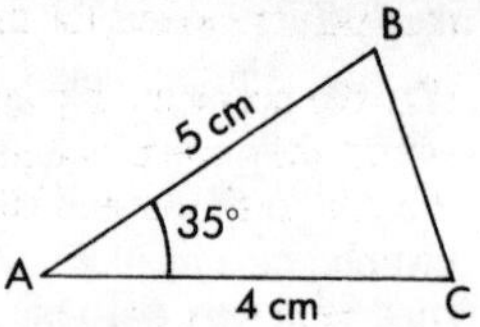

Fig C.22 Cosine rule

- If two sides and the included angle are known (as in Figure C.22), then the missing *side* can be found.
  Where the sides b and c are known, together with the angle A between them, then the cosine rule is:

$$a = \sqrt{(b^2 + c^2 - 2bc \cos A)}$$

$$\text{or} \quad b = \sqrt{(a^2 + c^2 - 2ac \cos B)}$$

$$\text{or} \quad c = \sqrt{(a^2 + b^2 - 2ab \cos C)}$$

COURSEWORK

Coursework should be an important part of your assessment and will take the form of different types of assignments. These will vary from group to group, and a much fuller description of what it all entails can be found in the Longman GCSE Coursework Guide for Mathematics.
Whatever the nature of the coursework, the assessment will be made in three main areas: practical, investigational and extended pieces of work.

COURSEWORK AREAS

Practical work

The assessment of practical work will include:

a) How well was the task planned? How was it carried out and how accurate were the results?
b) Was the task fully understood and was an appropriate use made of various items of equipment?
c) How well has the solution of the task been communicated? Have reasons been given for each stage of the solution?

Investigational work

The assessment of investigational work will include:

a) How well was the task planned? How was it carried out and how accurate were the results?
b) How much relevant information was obtained and used?
c) How well was the solution communicated? Could any valid conclusions be made about the task?

Extended work

The main points looked for in an *extended* piece of work are:

a) Was the problem fully understood and did the candidate show they knew where they were going?
b) Was the task planned into different set stages, and were all these stages completed?
c) How well was the task finally carried out? Was appropriate equipment made use of?
d) How well have the results been communicated?

The coursework tasks should be assessed at frequent, but appropriate, times during the course. Assessed coursework will help you be aware of how well you are doing. If you do have shortcomings you can then work at improving these. So do try to see your assessed coursework or at least talk to your teacher about it so that you can identify your strengths and weaknesses. weaknesses.
The final responsibility for your coursework lies with YOU, so you need to pay attention to:

- planning the work
- doing the work
- communicating the work done and the things found out.

CROSS SECTION

The cross section is the plane shape revealed by cutting a solid shape at right angles to its length (or height). Any shape that has the same cross section throughout its length (or height) is a **prism**.

CUBE

A cube is a three-dimensional shape with six square faces at right angles to each other.
But the term, *cube*, has other meanings:

- *To cube* is to multiply the same number by itself *twice*. For example the cube of 2 is $2 \times 2 \times 2$ which is 8. The *volume of a cube* is found by cubing the length of any edge.
- The *cube root* of a number, say P, is that number which, when multiplied by itself twice, gives the number P. For example, the cube root of 27 is 3, since $3 \times 3 \times 3 = 27$. The shorthand for cube root is $\sqrt[3]{\ }$.

CUBIC EQUATION

A cubic equation is one which has a cube as the highest power, e.g.
$x^3 + 2x^2 - 5 = 0$

CUBOID

A cuboid is a three-dimensional shape that has each opposite face congruent and at right angles to each other. It is also often called a *rectangular block*. Its volume is found by multiplying length by width by height.

CUMULATIVE FREQUENCY

This is sometimes known as *running totals*. They can be quite useful to show the spread of data and to find the **quartiles** of a distribution.

For example, the number of passengers carried on the buses on one particular route during one week was summarised as shown in Figure C.23.

Number of passengers	*Frequency*	*Cumulative frequency*
less than 10	12	12
11 – 20	23	35
21 – 30	28	63
31 – 40	28	91
41 – 50	9	100

Fig C.23 Cumulative frequency

When we draw the cumulative frequency graph (Fig C.24), notice the distinctive shape of the *cumulative frequency curve*; it is known as the **ogive** (in a dictionary 'ogive' means 'diagonal'). It can be used to find the **median** and the **quartiles.**

- **Exam Questions**

  The information in Figure C.25 shows the percentage distributions by age of teachers of mathematics and physical education in schools in 1984.

  a) Using the same axes and scales, draw cumulative frequency diagrams to represent this information. (You can assume that there were no teachers aged under 21 or aged 65 and over.)
  b) Estimate the difference between the median age of the mathematics teachers and the median age of the physical education teachers.
  c) Estimate the interquartile range for each distribution.
  d) What do you notice about your answers to parts b) and c)?
  e) Given that there were 47,900 mathematics teachers, estimate the number aged 45 and over.

(NEA;H)

	Aged under 30	*Aged 30–39*	*Aged 40–49*	*Aged 50 and over*
Mathematics	*17*	*36*	*29*	*18*
Physical education	*28*	*39*	*21*	*12*

Fig C.25

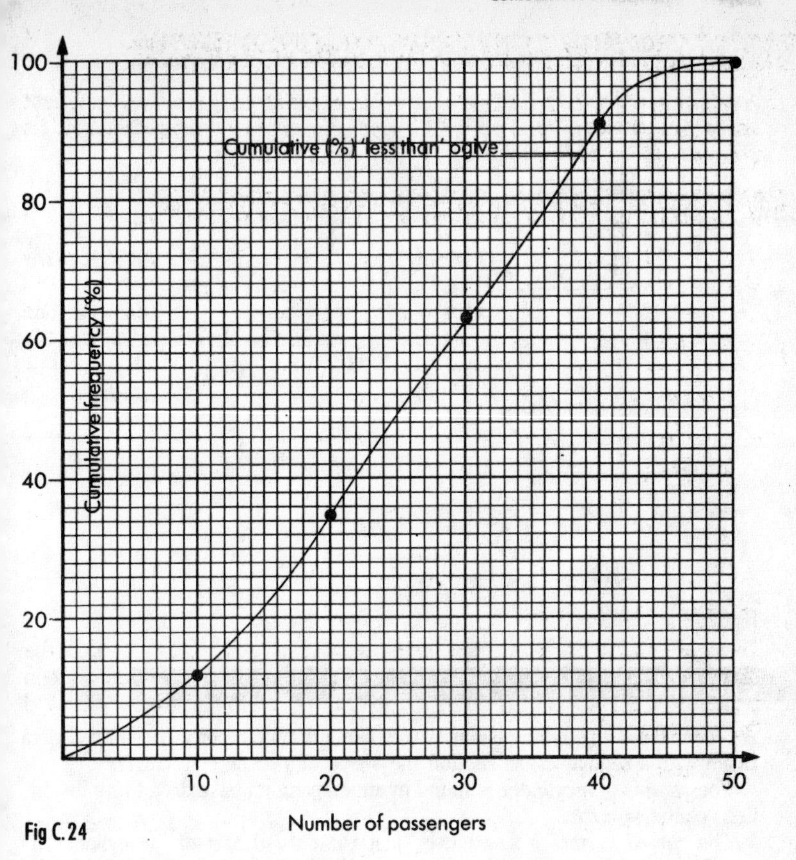

Fig C.24

■ **Solution**

a) see Figure C.26

b) median maths = 39
median P.E. = 35

c) IQ = UQ − LQ
where IQ = Interquartile Range
UQ = Upper Quartile
LQ = Lower Quartile
maths IQ = 45 − 32 = 13
P.E. IQ = 42 − 29 = 13

d) that although the medians are different, the interquartile ranges are the same.

e) 45 is the Upper Quartile (U.Q.); hence the number above this age will be one quarter of 47,900, which is 11,975. An approximation at 12000 would be a good number to use here.

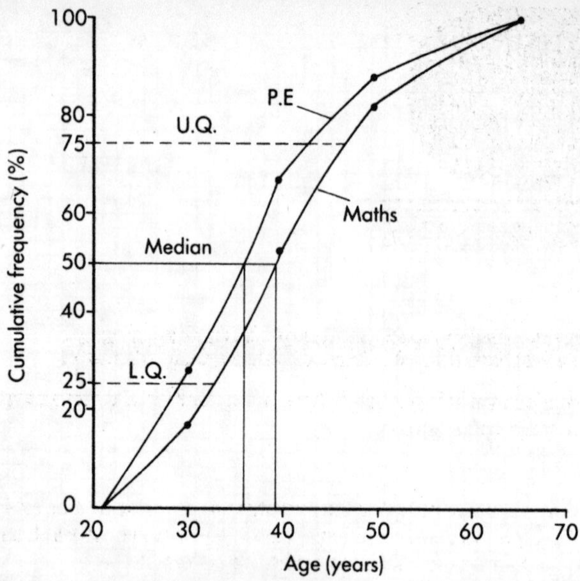

Fig C.26

CYLINDER

A cylinder is a three-dimensional shape like a drain pipe or a cocoa tin. It is a **prism** with a regular cross section the shape of a circle (Fig C.27).

The *volume* of a cylinder is found by multiplying its base area to its height, i.e. volume $= \pi r^2 h$.

The *curved surface area* (that is, just the curved part of the cylinder) is found by multiplying the circumference of the base to its height, i.e. curved surface area $= \pi D h$ (where D is diameter, $= 2r$).

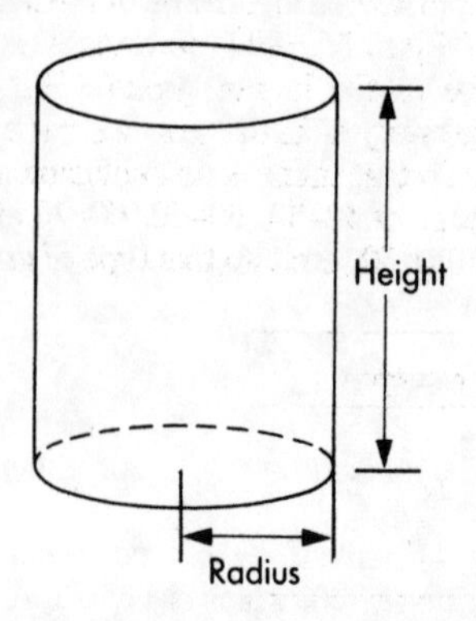

Fig C.27 Cylinder

DATA

Data is information available for use. There are a number of different types of data that you could come across.

Discrete data

This is data that can be identified on its own by a *single number*. For example, the number of goals scored by various teams, the number of children in a family or the number of records in a teenager's collection.

Continuous data

This is data that *cannot* be measured exactly, and so is given to a rounded off amount. For example, a length of time or the height of trees.

Grouped data

Often we are given non-precise information. This is the case in Figure D.1, where a survey was carried out to find the average weight of a class of children. The data has been recorded to the nearest kg and put into the groups as shown.

It is an essential feature of a *grouped frequency table* such as this that we know what the *group boundaries* are. Here, for example, the first group will take in weights from 40 kg to 50.4999 (note that the 50.5 would be **rounded off** to 51 and hence be in the higher group). But in the table it is more convenient to show the weights in the way we have done here.

Care has to be taken so that there is no confusion at group boundaries. For example, if we had groups of 40–50, 50–60, 60–70 etc., then it would not be clear what happens at 50 or 60, etc., so this type of grouping must be avoided.

Weight	*Frequency*
40 – 50	3
51 – 60	14
61 – 70	8
71 – 80	1

Fig D.1

DECAGON

A ten-sided **polygon**.

DECIMAL PLACES

Decimal places are to the right of the decimal point. So the number 4.782 has three places of decimal. Figure D.2 illustrates how **rounding off** is applied to decimal places.

Many mistakes are made in examination answer papers in rounding off decimal places.

	Decimal places		
Number	1	2	3
14.5638	14.6	14.56	14.564
0.8572	0.9	0.86	0.857
0.0295	0.0	0.03	0.030

Fig D.2

DECIMALS

A decimal is the name used to describe a number with a *decimal point* in it. This decimal point separates the whole numbers from the decimal fractions.

Recurring decimals

Some decimal numbers go on and on, following a particular pattern. For example 0.3333333 . . .

We would call this 0.3 *recurring* or use the mathematical notation $0.3\dot{3}$.

Some recurring decimals have more than one digit in the repeated pattern, e.g. 0.191919 . . . ; this would be represented as $0.1\dot{9}$.

There are a number of well know *vulgar fractions* that have recurring decimals as their equivalents. The way to find the decimal from a vulgar fraction is to divide the top number by the bottom number. Try this out with fractions like $\frac{1}{11}$, or $\frac{4}{9}$ or even $\frac{2}{7}$.

Terminating decimals

Any decimal number that actually *stops* at so many decimal places is called a *terminating decimal.* For example $\frac{1}{4}$ will stop at 0.25 and $\frac{3}{8}$ will stop at 0.375. These are just two examples of terminating decimals.

DEPENDENT VARIABLE

◀ Variable ▶

DEPOSIT

A deposit is the initial payment made on a hire purchase agreement. For example, a TV priced at £450 was offered on terms which include a deposit of £45 and a further 12 payments of £40.

The deposit is the first payment made. Usually it is the payment that has to be made before you can take the goods away with you.

DEPRESSION

The depression is the angle made with the horizontal while looking down. It is shown as angle A in Figure D.3.

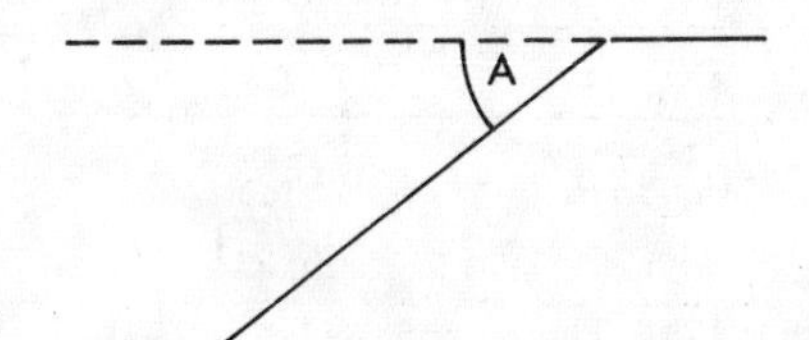

Fig D.3 Depression

DIAGONAL

A diagonal is a straight line that joins any two corners of a plane geometric shape.

DIALS

A dial is a round face with numbers on, usually looking like a clock but only going round from 0 to 9 so that the reading is a *decimal reading*. Some of the faces appear reversed and are read reversed. Look at the readings on the assorted dials in Figure D.4 to see how they may be used.

Fig D.4 Dials

- **Exam Questions**
  Figure D.5 shows the readings from a gas meter. Write down in figures each reading and find the difference between them.

(NEA;L)

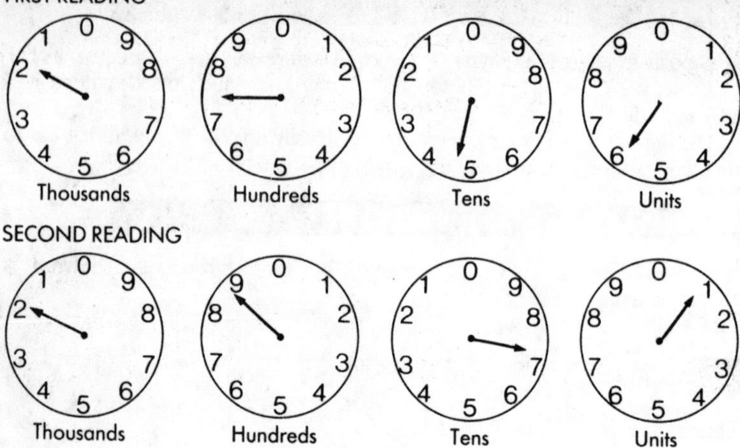

- **Solution**
  The first reading is 1746 The second reading is 1871
  So, the difference is 1871 – 1746 = 125

Fig D.5

DIAMETER

A straight line drawn from one side of the circumference of a circle to the other side, passing through the centre, is called a diameter. Any circle will have millions of diameters, all of the same length.

DICE

A dice is a regular three-dimensional shape, a **polyhedron**, with numbers on it. The most common sort of dice and certainly the type meant in any examination question (unless otherwise stated) is the six-sided cube with the numbers 1 to 6. These numbers are locked in such a way that the opposite faces add up to 7 each time.

A single dice is sometimes referred to as a *die*.

DIE

◀ Dice ▶

DIRECTED NUMBERS

Directed numbers is a term sometimes used to describe positive and negative arithmetic. Look at the number line in Figure D.6, and you will see a scale rather like that on a thermometer. There is a 0; then above the zero are positive numbers and below the zero are negative numbers. The negative numbers are also called *minus* numbers; for example 3 below the zero is usually referred to as minus 3, but written as −3.

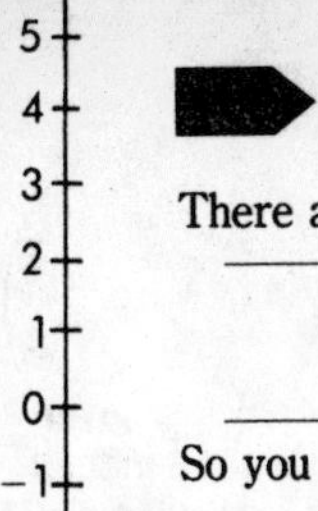

ADDING AND SUBTRACTING

There are a set of rules that you ought to know.

$+\ -$ is the same as $-$
$-\ -$ is the same as $+$

So you see that $4 + -3$ is the same as $4 - 3$, which is 1,
and $4 - -3$ is the same as $4 + 3$, which is 7.

For example:

$2 + 6 = 8$	$2 - 6 = -4$	$-2 + 6 = 4$	$-2 - 6 = -8$
$2 - -6 = 8$	$2 + -6 = -4$	$-2 - -6 = 4$	$-2 + -6 = -8$

Fig D.6

MULTIPLYING AND DIVIDING

There are a set of rules that you ought to know.

Rules for multiplying	+	×	+	=	+	*Signs the same*
	−	×	−	=	+	*answer:* +
	+	×	−	=	−	*Signs different*
	−	×	+	=	−	*answer:* −

An easy way to remember these rules is that when the *signs are the same* the answer is a *positive*, and when the *signs are different* then the answer is *negative*.

For example:

$2 \times 3 = 6$	$2 \times -3 = -6$	$6 \div 3 = 2$	$6 \div -3 = -2$
$-2 \times -3 = 6$	$-2 \times 3 = -6$	$-6 \div -3 = 2$	$-6 \div 3 = -2$

Many errors are made by candidates in this kind of arithmetic. These are mainly careless errors of getting the signs wrong, so do learn the rules.

- **Exam Questions**
  a) How many degrees colder is Londonderry at 6 a.m. than at midday?
  b) The temperature in Omagh rises by 6°C between 6 a.m. and midday. Give the temperature at midday. (NISEC;B)

Fig D.7

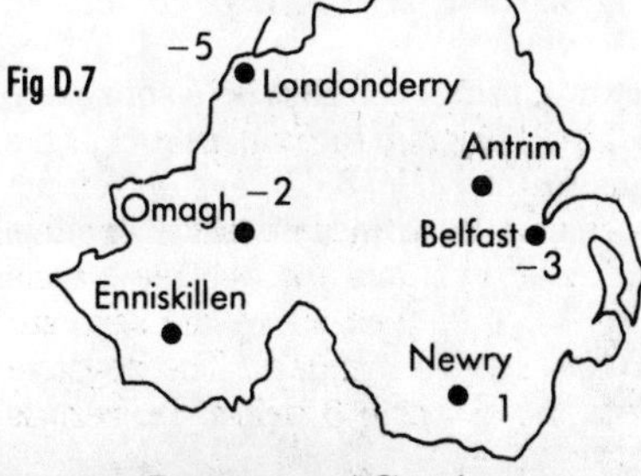

Temperature (°C) at 6 a.m.

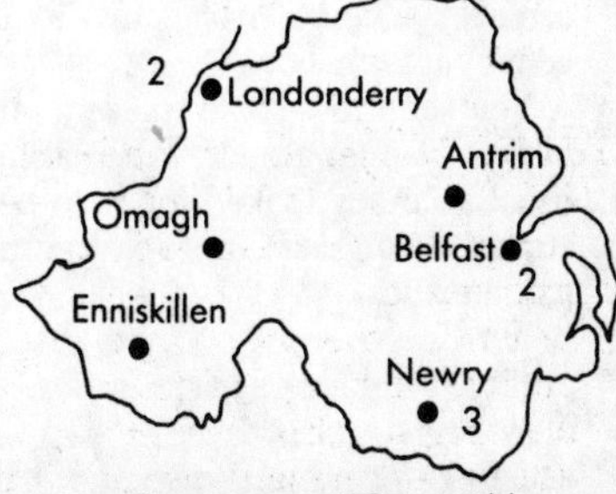

Temperature (°C) at midday

■ **Solution**

a) $2 - -5 = 7°C$
b) $-2 + 6 = 4°C$

DISCOUNT

A discount is a deduction from the usual amount paid. Often people are given a discount for paying cash for their goods or for being an employee or a member of a particular union. This discount is usually going to be a *percentage*, and always a percentage *reduction*. For example:

James as a paper boy in a newsagent's shop is allowed 5% discount on any goods bought over £1. How much will he pay for a game priced at £7.60?
The 5% discount on £7.60 will be $7.60 \times 5/100 = £0.38$
So James will pay £7.60 − £0.38 = £7.22

DISCRETE DATA

Discrete data is data that can be identified on its own by a single number. For example, the number of teeth you have or the number of televisions in your home.

◀ Data ▶

DIVIDING/DIVISION

◀ Directed numbers, Matrices ▶

DOMAIN

The domain of a **function** is the set of numbers that a function can work on. These numbers will **map** to the **range**, that is the set of numbers that the function takes the domain to.

For example in the function $f(x) = 3x^2$, the *domain* can be the whole set of numbers both positive and negative (and zero). However the *range* can only be positive numbers (and zero).

DRAWING

The type of drawing referred to in a GCSE mathematics syllabus is *geometrical drawing*. The main objective is drawing diagrams accurately or to scale to convey information.

You need to be able to use a protractor, a pair of compasses, a set square and, of course, a ruler. When asked to draw a diagram (or to **construct** it) the usual accuracy looked for is to be no more than 1 or 2 degrees out on the angles. Being inaccurate is the most common error made while drawing in mathematics.

EDGE

An edge is found on three-dimensional shapes where two faces meet, as in Figure E.1.

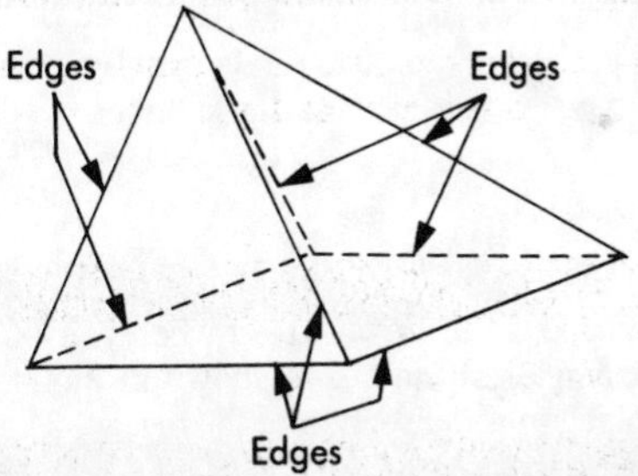

Fig E.1 Edge

ELEVATION

In GCSE mathematics you need to be aware of the different aspects of elevation.

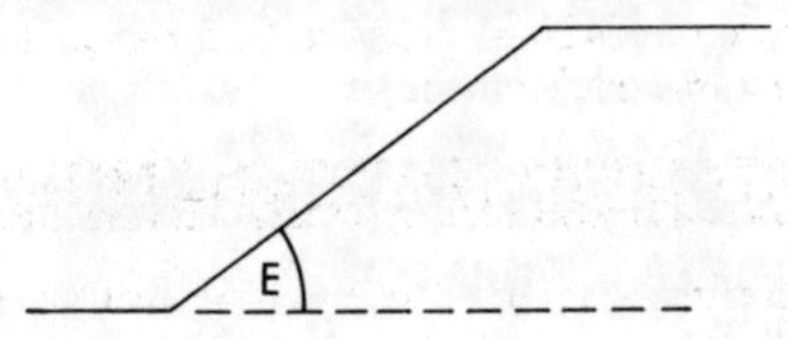

Fig E.2 Angle of elevation

Angle of elevation

The angle of elevation is the angle made with the horizontal while looking up, as in Figure E.2.

End elevation

The end elevation of a shape is the view you get when looking at the end of a shape. For example, in the shape shown in Figure E.3, the end elevation is found by looking in the direction of the arrow. The end elevation is drawn underneath the shape.

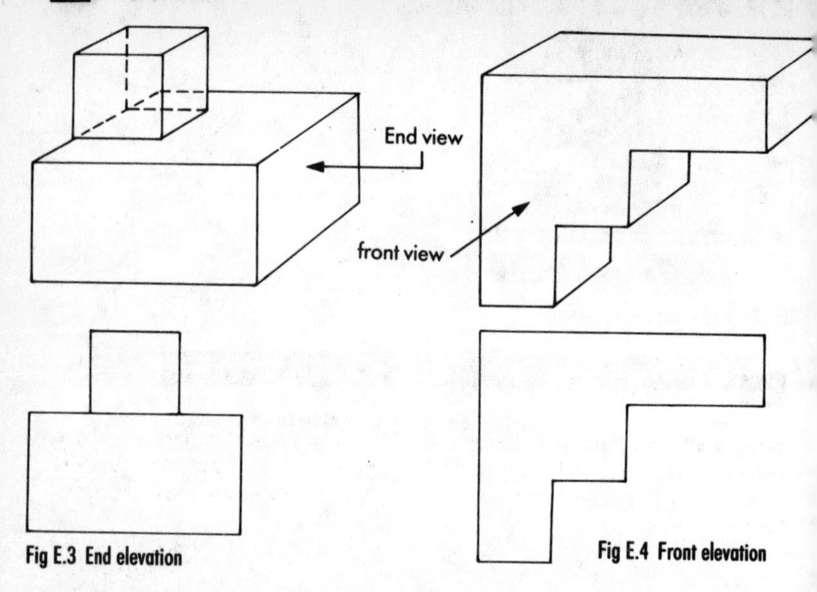

Fig E.3 End elevation

Fig E.4 Front elevation

Front elevation

The front elevation of a shape is the view you get when you look directly at the front of a shape. For example, in the shape shown in Figure E.4, the front elevation is found by looking in the direction of the arrow. The front elevation is drawn underneath the shape.

The most common errors are to draw an elevation with **Perspective**, that is to try and make it look like a three-dimensional drawing instead of a plane diagram in two dimensions only.

ENLARGEMENT

When all the respective dimensions of two shapes are in the *same ratio*, then each shape is an *enlargement* of the other.

GEOMETRIC ENLARGEMENT

The idea of an enlargement is to make a shape of different size (usually larger but it could be smaller), and in a specific place. You will be given a *scale factor* which tells you how many times bigger each line will be. You will also be given a *centre of enlargement* which will determine *where* the enlargement ends up.

For example, to enlarge the shaded shape in Figure E.5 with a scale factor of two from the centre of enlargement *, the distance from the centre of enlargement to each vertex in the shape is multiplied by two. This is shown in the diagram by the dotted lines. The enlarged shape will have all its dimensions multiplied by the scale factor, but will keep all angles the same. In other words, the two shapes will be **similar** shapes.

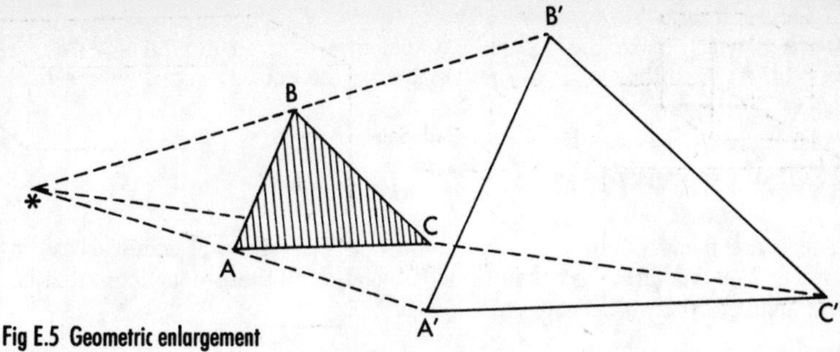

Fig E.5 Geometric enlargement

NEGATIVE ENLARGEMENT

When asked to draw an enlargement with a *negative* scale factor, then you need to draw the lines *back* through the centre of enlargement (the opposite way round to a positive scale factor that we described above). Then enlarge as before.

In Figure E.6 the shaded triangle has been enlarged with a scale factor of −2 with the origin as the centre of enlargement. The enlarged shape under *negative* enlargement will always end up *upside down*, but will still be a **similar** shape to the original.

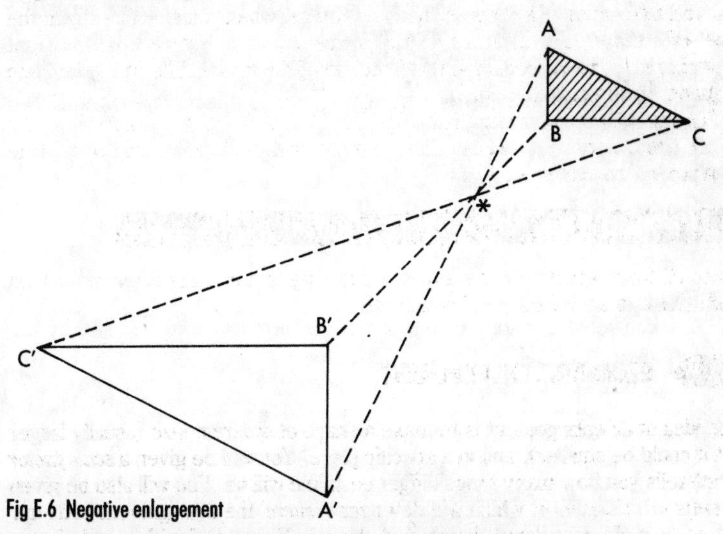

Fig E.6 Negative enlargement

EQUATIONS

Equations are mathematical statements that one expression is equal to another. In other words two (or more) quantities are joined by an equal sign.

For example, $4x + 2 = 2x + 10$ is an equation.

The *solution* of an equation refers to the value (or values) of the variable (here x) which make the equals sign hold true. Our solution in this case is $x = 4$. At any other value of the variable, x, the equality sign does not hold true.

TYPES OF EQUATION

There are many different types of equation that you will come across in GCSE. You will often be asked to find a solution to these equations, that is, the numbers that make both sides equal.

Linear equations

Linear equations involve single variables of power 1. They contain no expressions such as x^2, x^3, $1/x$, xy etc.

For example, $x + y = 6$ $5x - 3 = 2y$ $4(2x + y + 4z) = 3$

These can usually be solved by moving the equation around using the well known phrase – 'If it's doing what it's doing to the rest of that side then you can move it from one side of the equation to the other and make it do the opposite job.'

Quadratic equations

A quadratic equation is an equation involving a variable with the power 2, e.g. $x^2 + 5x + 6 = 0$. Here the *highest* power is a two. To solve quadratic equations you can use one of three methods:

1 Factorise

Quadratic factorisation is where we put the quadratic back into two brackets and solve those brackets.
For example, solve $x^2 + x - 12 = 0$
Factorising into two brackets gives us $(x + 4)(x - 3) = 0$
Hence $(x + 4) = 0$ or $(x - 3) = 0$
hence $x = -4$ or $x = 3$.
The solution is $x = -4$ and 3.

2 Formula

You can solve a quadratic equation of the form $ax^2 + bx + c = 0$ by the formula:

$$x = \frac{-b \pm \sqrt{b^2 - 4ac}}{2a}$$

For example, solve $2x^2 + 5x - 3 = 0$
Use the formula where $a = 2$ $b = 5$ $c = -3$

then $x = \dfrac{-5 \pm \sqrt{25 - -24}}{4}$

$x = \dfrac{-5 \pm \sqrt{49}}{4}$

then $x = \dfrac{-5 \pm 7}{4}$

hence $x = -12/4$ and $2/4$

so $x = -3$ and 0.5

3 Completing the square

You can solve any quadratic equation by making a square and rooting.

For example, solve $x^2 - 8x + 12 = 0$

Find a square as near as possible that fits the left hand side. For instance $x - 4^2$ which is $x^2 - 8x + 16$.

Our equation can now be rewritten as:

$$x^2 - 8x + 16 = 16 - 12 = 4$$
$$(x - 4)^2 = 4$$
$$(x - 4) = 2 \text{ or } -2$$

hence $x = 2 + 4 = 6$ or $x = -2 + 4 = 2$

so $x = 6$ and 2

Any of the above methods can be used to solve your quadratic equation, but do use the method that you are familiar with and are confident about using.

◀ Formula, Identities, Variables ▶

Fractional equations

Fractional equations of the type $\frac{x}{3} + \frac{x-2}{4} = 2$ can be solved by changing both fractions to *equivalent fractions* with the same bottom number (denominator). Then carry on using normal algebraic techniques.

For example, solve the equation $\frac{x}{3} + \frac{x-2}{4} = 3$

Change both fractions to a 12 on the bottom hence $\frac{4x}{12} + \frac{3x-6}{12} = 3$

$$4x + 3x - 6 = 36$$
$$7x = 42$$
$$x = 6$$

Simultaneous equations

Simultaneous equations are where two equations both need solving at the same time. These equations are usually linear but not necessarily so, and one could well be quadratic. There are two common methods for solving simultaneous equations:

1 Substitution method

The technique is to *substitute* one equation into the other and so eliminate one of the variables. Then solve for the variable left and substitute back to find the value of the eliminated variable.

For example, solve (i) $x + 2y = 13$ . . . .

(ii) $4x - y = 7$. . . .

Substitute (i) . . . $x = 13 - 2y$ into (ii) to give $4(13 - 2y) - y = 7$

Now we only have one unknown variable, y, and one equation.

$$52 - 8y - y = 7$$
$$52 - 7 = 9y$$

hence $45 = 9y$

so $y = 5$

Now substitute this back into equation (i) to give us $x + 10 = 13$, so $x = 3$.

The solution of the simultaneous equations is $x = 3$ and $y = 5$

2 Elimination method
The technique is to *eliminate* one variable by adding or subtracting the two equations. Then proceed as above.
For example, solve $3x - y = 9$ . . . . (i)
$5x + y = 23$ . . . . (ii)

Add (i) to (ii) to eliminate y, then $8x = 32$, hence $x = 4$.
Substituting into (i), then $12 - y = 9$, and $3 = y$.
So the solution of the simultaneous equation is $x = 4$ and $y = 3$.

The most common error to be made here by candidates in examinations is not to *check* their answers. A simple and effective technique is to check that your solution works by substituting it into the equation you did not substitute into before. Doing this at least allows you a check to see if you could be wrong. If you are, then you can have another go if you have time.

■ **Exam Questions**
Solve the simultaneous equations $4x - y = 3$ . . . . (i)
$3x + 2y = 16$ . . . . (ii)

(HIGHER)

■ **Solution**
Here we use the *elimination* method, though the substitution method would be equally acceptable.
To eliminate y we need to double the whole of equation (i) to give

$8x - 2y = 6$
$3x + 2y = 16$

Now we can add the equations to eliminate y, to give $11x = 22$, making $x = 2$, which is substituted into equation (i) – being the simplest – to give $8 - y = 3$, or $y = 5$. So the final solution is $x = 2$, $y = 5$.

EQUIVALENT

Two expressions that are equivalent will have the same value but could well *appear* different.

Equivalent fractions

These are *fractions* that have the same value but which look different. Any two equivalent fractions will have the same **decimal fraction** and will cancel down to the same **vulgar fraction**.

For example, $\frac{1}{2}$, $\frac{3}{6}$, $\frac{6}{12}$, $\frac{7}{14}$, $\frac{2}{4}$, etc. are all equivalent fractions that cancel down to 0.5 or $\frac{1}{2}$.

ESTIMATION

An estimation is a calculated guess at some length, weight or other amount. We *estimate* by referring to things we already know. For example, it will help in estimating the weight of a book if we can already recognise a particular weight, like a bag of sugar, which represents 1 kg. We can then ask ourselves whether the book is twice that weight, or the same, or even smaller, etc.

To be good at estimating depends partly on experience and partly on a knowledge of standard weights and measures.

EXCHANGE RATES

The pound abroad (1st April 1989)

Austria	Sch 19.66	Hong Kong	HK$ 13.66
Belgium	Fr 68.59	Ireland	I£ 1.225
Canada	C$ 2.08	Italy	L 2393
Denmark	Kr 12.7	Norway	NKr 11.81
France	Fr 11.1	Spain	Pes 207
Germany	DM 3.03	Switzerland	Fr 2.82
Greece	Drc 210.5	USA	$1.715
Holland	Gld 3.42		

Fig E.7 Exchange rates

Figure E.7 shows the exchange rates for the British pound on April 1st, 1989. The table indicates the amount of each foreign currency you would have received for £1 on that day. These rates do change day by day, so any table soon becomes out of date. To exchange *back* into British pounds, you will need another table. This will have different rates since the money exchangers take their cut! So you may well exchange £50 for say 555 French francs, but then if you try to exchange this back into pounds you would probably only get about £45.

EXPAND

To expand within mathematics will usually mean to multiply out the brackets. For example expand $5(3x + 4y)$, will mean multiply the bracket by 5 to give $15x + 20y$.

EXPECTATION

This often refers to that part of **probability** where we try to work out the *expectation* of a certain result. This is found out by multiplying the *probability of an event* by the *number of trials* for that event. For example:

A doctor found by random trial that out of 50 of his patients 17 of them had back trouble. How many of his 2000 patients are likely to suffer back trouble?

The probability of any *one* patient at random having back trouble is $^{17}/_{50}$.

So the *expectation* of the total number of his 2000 patients having back trouble will be $^{17}/_{50} \times 2000$, which is 680. So the doctor can expect around 680 of his patients to have back trouble.

EXPONENT

An exponent is the power to which a term has been raised. For example in 10^3 the exponent is 3. You will see this on your scientific calculator in two ways:

1 The E that you sometimes see on the calculator display. For example 4.56 E 9 is the calculator shorthand for *standard form,* where the number is actually 4.56×10^9 or 4,560,000,000.

2 The button that is marked either EXP or E or even EE (check to see which it is) on your calculator. You use this to put a *standard form* number into the calculator yourself. For example, to put in the standard form number 8.5×10^2, you would key in 8.5 EXP 2. Try it out.

EXTENDED WORK

This is part of your coursework.

◀ Coursework ▶

FACE

A face is a surface of a three-dimensional figure bounded by edges. **Edges** are where two faces meet.

FACTOR

The factors of a whole number N are the whole numbers that will divide into N exactly. For example, the factors of 16 are 1,2,4,8 and 16.

- **Exam Questions**

a) List the set of factors of
   i) 48
   ii) 72
b) List the common factors of 48 and 72
c) i) List the prime factors of 48
   ii) Express 48 as a product of primes

(NEA;I)

- **Solution**

a) i) 1, 2, 3, 4, 6, 8, 12, 16, 24 and 48
   ii) 1, 2, 3, 4, 6, 8, 9, 12, 18, 24, 36 and 72
b) 1, 2, 3, 4, 6, 8, 12, 24
c) i) 2, 3
   ii) $2^4 \times 3$

FACTORISATION/FACTORISE

To factorise means to separate an expression into the parts that will multiply together to give that expression. The two (or more) parts are usually connected by brackets. For example:

$2p + pt$ can be factorised to give $p\,(2 + t)$
$5ab - ab^2$ can be factorised to give $ab\,(5 - b)$

Quadratic factorisation

This means putting a *quadratic expression* back into its brackets, if at all possible. It is helpful when factorising to first consider what the *signs* may be:

1 When the last sign in the quadratic $ax^2 + bx + c$ is *positive*, then both signs in the brackets are the *same* as the *first sign* in the quadratic. For example:

$$x^2 + 5x + 4 = (\quad + \quad)(\quad + \quad)$$
and $$x^2 - 5x + 4 = (\quad - \quad)(\quad - \quad)$$

2 When the last sign in the quadratic $ax^2 + bx - c$ is *negative*, then the signs in the brackets are *different*. For example:

$$x^2 + 5x - 4 = (\quad + \quad)(\quad - \quad)$$

Once you've sorted out the *signs*, then you need to look at the *numbers*. Follow through these two examples to see how to do this

- **Example 1**
Factorise $x^2 + 6x + 8$
By looking at the *signs* we see that the brackets both contain the *same* sign which will be a +. We then see that the *end numbers* in each bracket must multiply together to give 8. This could be 1×8 or 2×4. But the combination must add together to give 6, so this will be the 2×4 combination.
Hence the factorisation is $(x + 2)(x + 4)$.

- **Example 2**
Factorise $x^2 + x - 12$
By looking at the *signs* we can see that both the brackets will have a *different* sign. We then see that the *end numbers* will multiply together to give a 12. This could be 1×12 or 2×6 or 3×4. But the combination must give a difference of 1, so this will be the 3×4 combination. The larger number needs to be positive since the quadratic has $+x$ in the middle.
Hence the factorisation is $(x + 4)(x - 3)$.

- **Exam Questions**
Factorise completely $6xy^2 - 12x^2y$

- **Solution**
$6xy(y - 2x)$

FINITE

A finite number is a countable number. If a number is too big to be counted, then it is called **infinite**. For example, the number of grains of sand in the world. Although this last example is a fixed number at any one point in time, it is just impossible to count them all since there are so many. So we would use the term infinite in this case for finite and infinite sets.

◀ Sets ▶

FLOW CHARTS

Flow charts are charts which you 'flow' through. In other words you start at the point where the chart tells you to, then you work your way through this

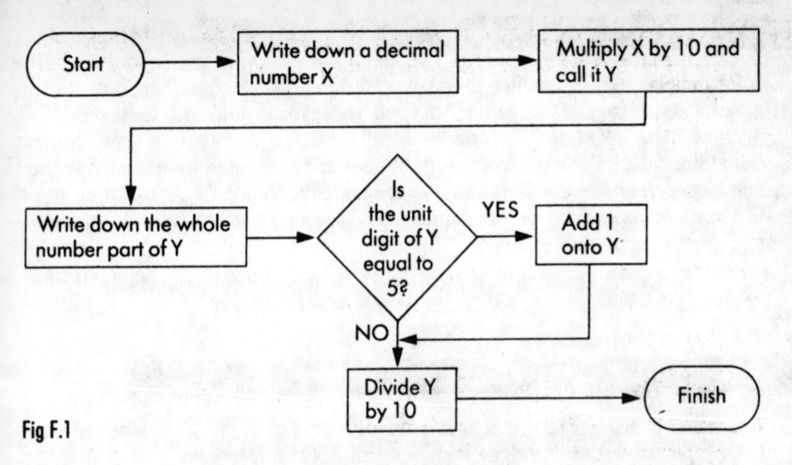

Fig F.1

chart until you come to the end. For example; follow through the flow chart in Figure F.1. This will round your number off to 1 decimal place.

- **Exam Questions**

  Abdul buys a ladder which costs £65 from a store 5 miles from his house. Use the flow chart in Figure F.2 to find the delivery charge.

  (SEG; I)

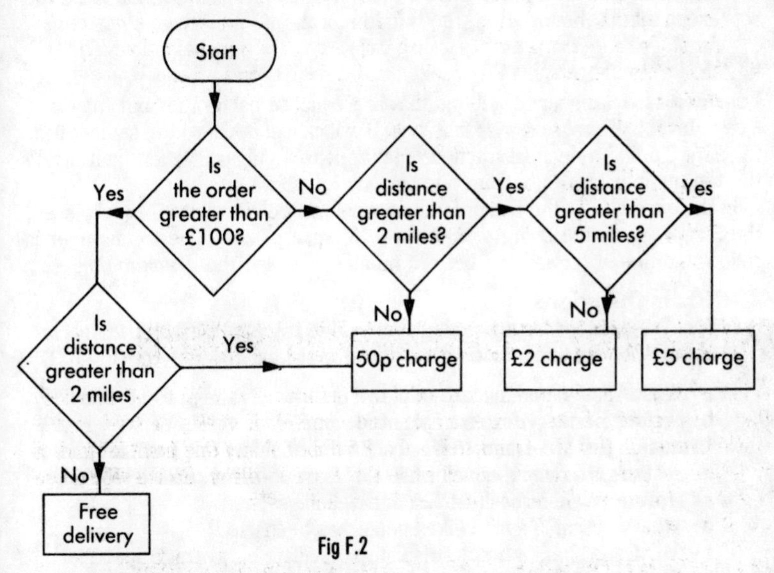

Fig F.2

- **Solution**

Follow through the chart correctly and you should end up with a £2 charge.

FORMULA

A formula is a set procedure to follow in order to work something out. It is a general expression that can be applied to several different values of the quantities in question. A formula can be in the form of a **flow chart**. Sometimes it can be a *sentence*, e.g. 'Wages are calculated by multiplying the number of years worked by £10 and then adding on £45.'

Most commonly, however, formula is expressed algebraically. For example:

Wages in £s = $10x + 45$, where x = number of years worked.

FRACTION

A fraction is an expression which contains a *part* of a whole. There are two main types of fraction: vulgar fractions and decimal fractions.

Vulgar fractions

A vulgar fraction is expressed using two whole numbers, one above the other. For instance $\frac{3}{5}$, $\frac{4}{6}$, $\frac{12}{17}$ and so on.

Decimal fractions

A decimal fraction is one that uses a decimal point and has decimal places to represent tenths, hundredths, thousandths, etc.

Equivalent fractions

Equivalent fractions are fractions that are equal to each other but may well *look* different. Equivalent vulgar fractions will cancel down to the same vulgar fraction. To find a decimal fraction equivalent to a vulgar fraction, just divide the top number by the bottom number.

In the example $\frac{3}{5}$, the 5 represents a whole unit divided into 5 equal pieces; the 3 represents the using of 3 of these 5 equal pieces. The top number is called the *numerator* and the bottom number is called the *denominator*.

- **Exam Questions**

Mr. Dane leaves his home near Sheffield to travel to Birmingham by car. Before he sets off he looks at his petrol gauge and mileometer (Fig F.3).

a) What fraction of a full tank of petrol is shown by the petrol gauge?
b) Write your answer to part a) as a decimal.

Later that day Mr. Dane arrives back in Sheffield and notices that his petrol tank is now one-quarter full and his mileometer reads 35468.

c) How many miles has Mr. Dane travelled?
d) What fraction of a full tank of petrol has been used?
e) Mr. Dane spent 4 hours travelling. Calculate his average speed for the journey. Give your answer to the nearest mile per hour.

(NEA;I)

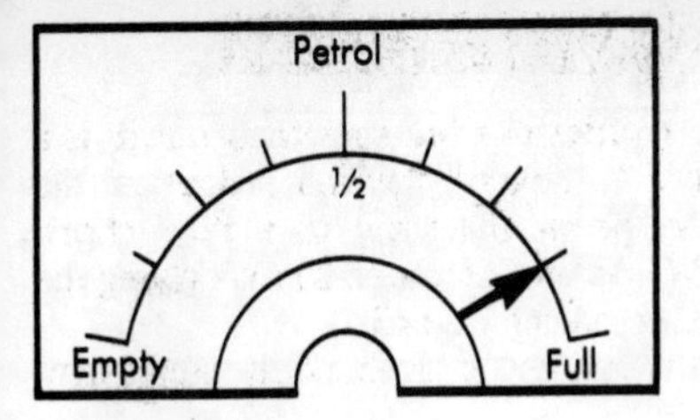

3	5	2	8	9

Miles

Fig F.3

■ **Solution**

a) $\frac{7}{8}$
b) 0.875
c) 35468 − 35289 = 179 miles
d) $\frac{7}{8} - \frac{1}{4} = \frac{7}{8} - \frac{2}{8} = \frac{5}{8}$
e) 179 ÷ 4 = 44.75
 = 45 mph.

FREQUENCY

Frequency is the number of times some defined event occurs. This can be found by the use of a **tally chart**. Frequency can be represented by the use of **bar charts, pictograms, pie charts** and **histograms.**

Grouped frequency

◀ Grouped data under data ▶

Cumulative frequency

Cumulative frequency is sometimes known as *running totals.* They can be most useful in finding **medians** and **quartiles.**
◀ Cumulative ▶

Frequency density

This is used as the vertical axis on a **histogram**. Frequency density is

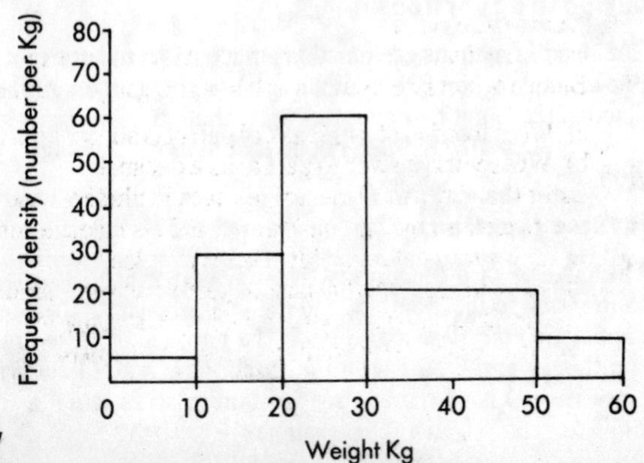

Fig F.4 Frequency density

Box weight	*f.d.* × *width*	*Frequency*
0 – 10	5 × 10	= 50
11 – 20	30 × 10	= 300
21 – 30	60 × 10	= 600
31 – 50	20 × 20	= 400
51 – 60	10 × 10	= 100

Fig F.5

abbreviated as f.d. The *frequency* that any block represents on a histogram is given by the *area* of the block. This is found by multiplying the width of the block by its height as given on the frequency density axis.

For example in the histogram in Figure F.4 the weight of an average-sized box handled by a delivery service is between 10 and 30 kg. The histogram illustrates the number of boxes handled in one particular week. From the histogram we can build up the table in Figure F.5 to illustrate the *frequency* of each group.

FUNCTIONS

A function is a mapping from one *set* to another where each element in the initial set, called the **domain**, will map to one, and only one, element in the **image set** (or **range**).

A function is another way of writing an algebraic formula. We generally use the notation of $f(x)$ for the image of x under the function f. For example, where the function $f(x) = 4x + 3$, then $f(2) = 11$, since substituting $x = 2$ into $4x + 3$ gives us 11. In other words, it is another way of writing $y = 4x + 3$ where we talk about $f(x)$ instead of y. Another way of expressing the function f is as $f{:}x \rightarrow 4x + 3$.

TYPES OF FUNCTION

Composite functions

Composite functions are functions made up to two (or more) simple functions. For example when $f(x) = 3x$ and $g(x) = x + 1$, then $gf(x)$ which is a *composite* function, is found by doing $f(x)$ first, then $g(x)$. So here $gf(2) = g(6) = 7$.

But $fg(2) = f(3) = 9$, so notice that in most cases $fg(x)$ will *not* be equal to $gf(x)$.

Inverse functions

The inverse of a function is that function which returns numbers from the **range** back to the **domain**, and the images back to the original elements. It is the function *the other way round.* The notation of an inverse function is f^{-1}. To find this inverse you have to go back. One way to always find an inverse (if there is one) is to change the function to an *equation* and then to *change the subject* of the formula. For example:

Find the inverse of the function $f: x \to 3x - 1$

- write the function as an equation... $y = 3x - 1$
- change the subject to give $x = \frac{y + 1}{3}$
- hence the inverse function is $f^{-1}: x \to \frac{x + 1}{3}$

◀ Self inverses ▶

- **Exam Questions**

Given that $f(x) = x + 1$, $g(x) = 4x - 3$ and $h(x) = x^2 + 1$,

a) Write down

i) $fg(x)$

ii) $hf(x)$

iii) $k(x)$, where k is the inverse of g

b) Find the set of values of x for which $g(x) > f(x)$

c) Explain why $hf(x) > 0$ for all x

d) For what value of x does the function $hf(x)$ have its minimum value?

(NEA;H)

- **Solution**

a) i) $fg(x) = f(4x - 3) = 4x - 3 + 1 = 4x - 2$

ii) $hf(x) = h(x + 1) = (x + 1)^2 + 1 = x^2 + 2x + 2$

iii) $g: x \xrightarrow{\times 4} 4x \xrightarrow{-3} 4x - 3$

$\frac{x + 3}{4} \xleftarrow[\div 4]{} x + 3 \xleftarrow[+3]{} x$

hence $g^{-1}(x) = \frac{x + 3}{4}$

b) $4x - 3 > x + 1$

$3x > 4$

$x > 4/3$

c) $hf(x) = (x + 1)^2 + 1$

since $(x + 1)^2$ is always positive, then so too must $(x + 1)^2 + 1$

hence $hf(x) > 0$ for all x.

d) the smallest value $(x + 1)^2$ can be is 0, when $x = -1$;

hence the minimum value will be, when $x = -1$

GENERALISE

To generalise means to find out some *pattern* in a situation and to express this in general terms using an algebraic expression. For example, in the sequence 3,6,9... we can see that the first number is 1×3, the second number is 2×3, the third number is 3×3, and so on. So the nth number in the sequence will be $n \times 3$ or $3n$.

- **Exam Question**

Alison often called in at the Rose Cottage Café with friends for a pot of tea. During one summer she noted that

a pot of tea for 2 people cost 80p,
a pot of tea for 3 people cost 95p
a pot of tea for 4 people cost £1.10.

a) What would be the cost of a pot of tea for
   i) 5 people?
   ii) 1 person?
b) Write down the formula that the Rose Cottage Café uses to calculate the cost of a pot of tea.

(NEA; I)

- **Solution**

Notice the list goes up 15p each time, hence

a) i) £1.10 + 15p = £1.25
   ii) 80p − 15p = 65p
b) Cost = 50p + $15n$, where n is the number of people, or
   cost = 65p + $15n$, where n is the number of extra people.

GEOMETRY

Geometry can be defined as the study of solid shapes, surfaces, curves, lines and points in space. There are many geometrical facts to remember, particularly those concerning triangles, polygons and circles.

Transformation geometry

Transformation geometry looks at how shapes change position and size according to certain rules that we call **transformations.** The common transformations at GCSE level are **translations, rotations, reflections** and **enlargements.** There are still others, such as **stretches** and **shears,** that also come into some syllabuses.

GRADIENT

The gradient of a line is its *steepness*; the bigger the gradient the steeper it is to go up the hill. A *negative* gradient means going downhill.

To be precise, the gradient of a straight line is the vertical distance divided by the horizontal distance between any two points on that straight line.

MEASURING GRADIENTS

Gradient on a straight line

In general, if we have a *straight line,* as in Figure G.1, and choose two points P (x_1, y_1) and Q (x_2, y_2) that both lie on the line, then the gradient is given by

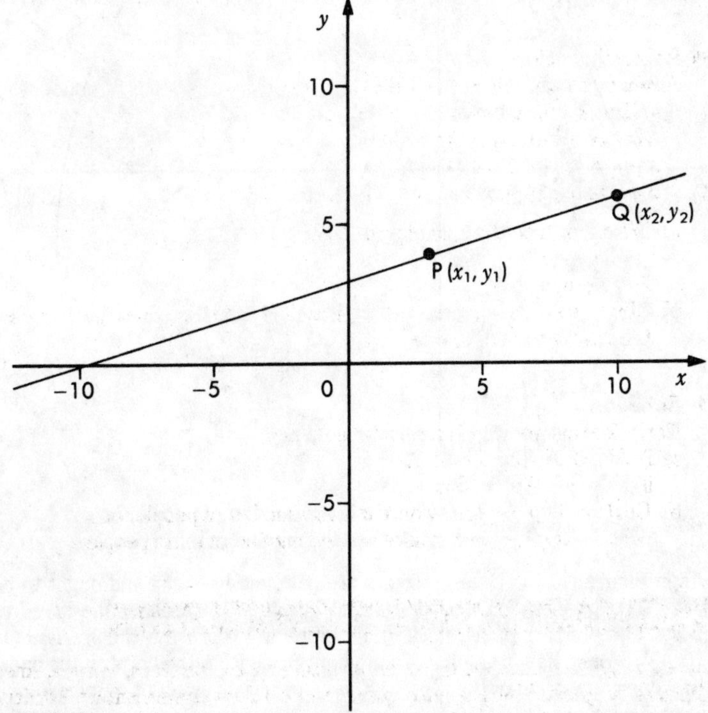

Fig G.1 Gradient

$(y_2 - y_1)/(x_2 - x_1)$. This will always automatically get the sign correct, which is + ve uphill and − ve downhill.

Gradient on a curve

To find the gradient on a *curve*, you must draw a **tangent** to the curve at the point in question. You must then find the gradient of the tangent . Of course the gradient keeps changing on a curve.

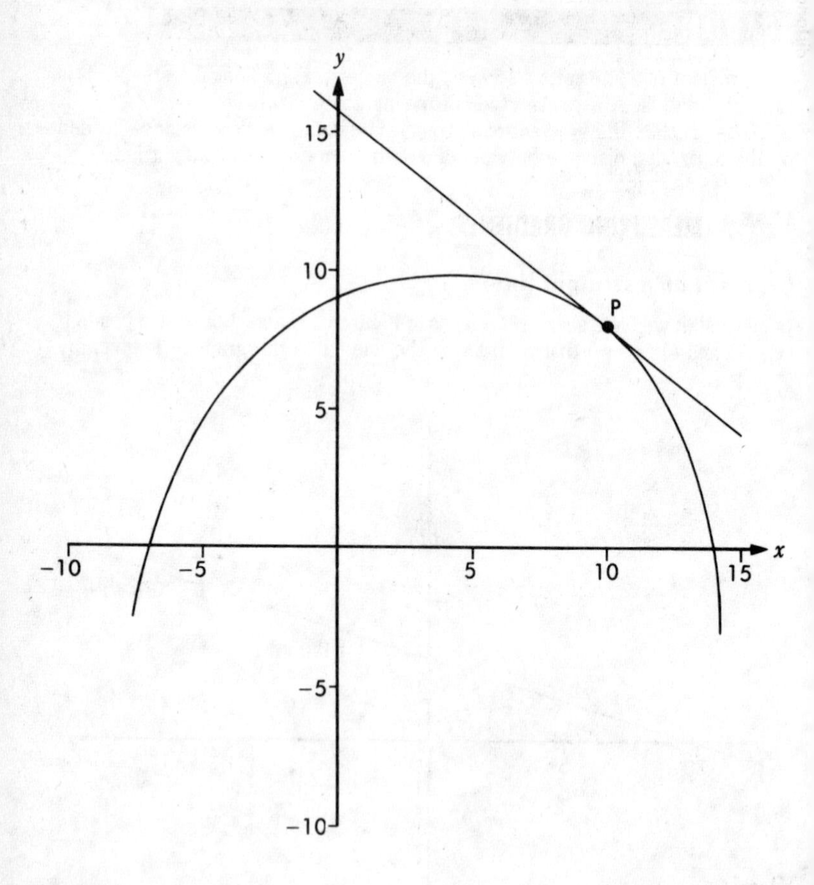

Fig G.2 Gradient on a curve

As in Figure G.2, certain gradients are particularly useful and ought to be known. The gradient on a distance/time graph will represent the *speed* or *velocity*. The gradient on a velocity/time graph will represent the *acceleration* (Fig G.3). Notice how the *units* of a gradient come from the units on the axes.

Any **linear** equation of the type $y = mx + c$ has a gradient m, where m is the coefficient of x (Fig G.4).

Fig G.3 Gradient a)

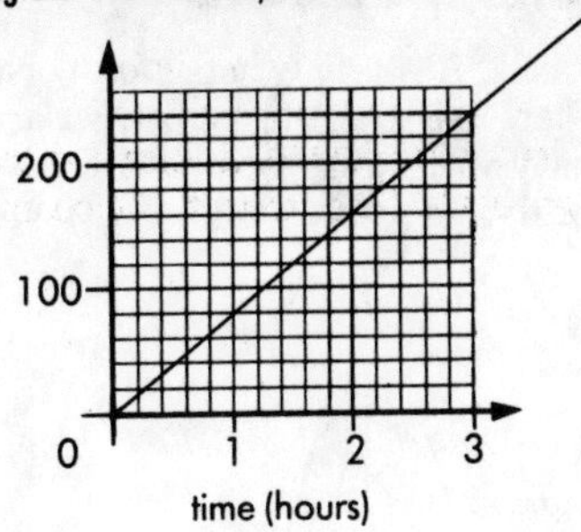

b)

150
100
50
Velocity Km/h
0
1
2
3
time (hours)

Velocity $= \dfrac{240\text{km}}{3\text{ hrs}} = 80$ km/h

Acceleration $= \dfrac{-125}{2.4} = -52.1$ km/h^2

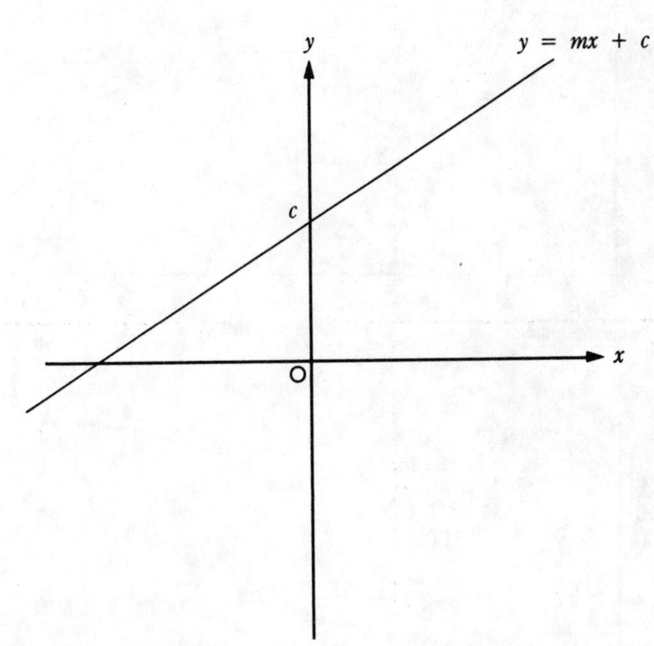

Fig G.4

- **Exam Question**

  A hill whose gradient is 1 in 5 can be said to have a gradient of 20%. Express a gradient of 1 in 7 as a percentage correct to the nearest integer.

  (NEA; H)

- **Solution**

  $\frac{1}{7} \times 100 = 14\%$ (to the nearest integer)

GRAPHS

A graph is a visual picture of information or data. It is usually in the form of a set of co-ordinates on a *cartesian grid*, where the axes are perpendicular. **Equations** have graphs to represent the many values that satisfy that equation; it is usually one of these that we refer to as a graph.

Fig G.5 Travel graphs

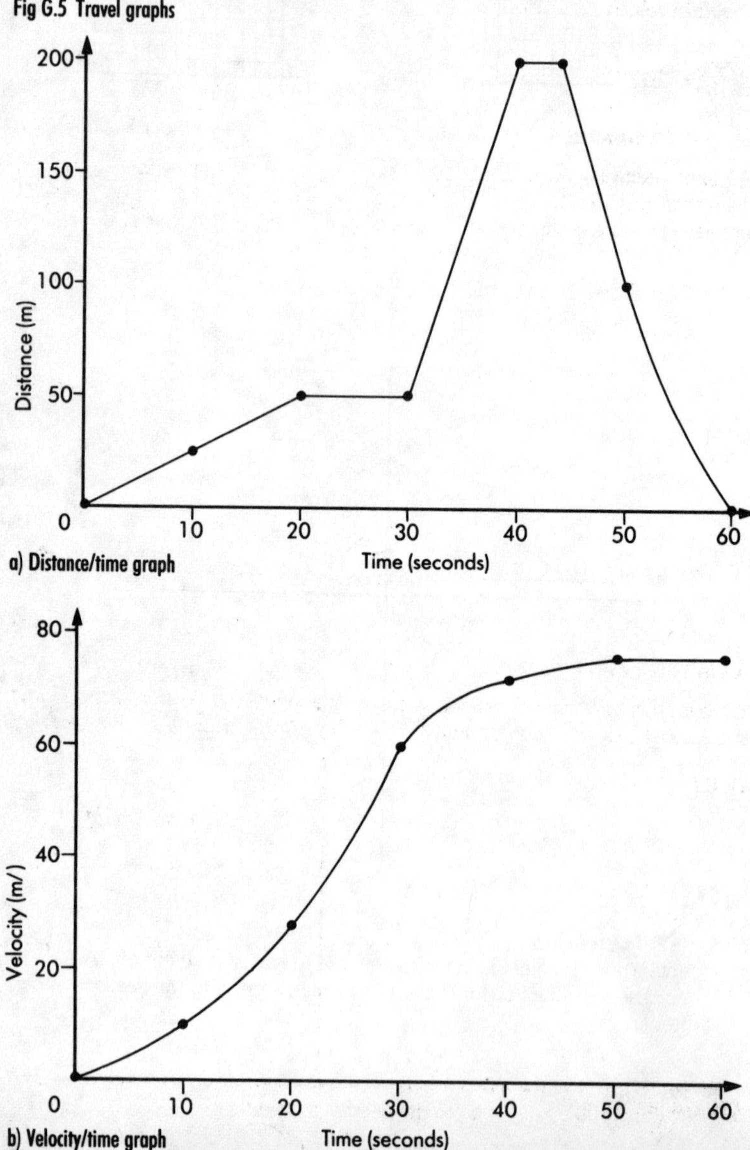

TRAVEL GRAPHS

These are graphs that represent a journey of some distance over some time or of some velocity over time. The two diagrams in Figure G.5 are both travel graphs that represent a journey over one minute:

a) is a distance/time graph
b) is a velocity/time graph.

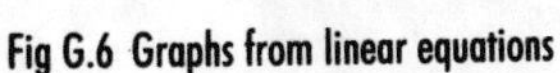

GRAPHS FROM EQUATIONS

Linear equations

All linear equations are of the form $y = mx + c$, where m and c are two constants (Fig G.6). They will be represented by a *straight line graph* of gradient m passing through the y axis at $y = c$ (the y axis *intercept*).

Fig G.6 Graphs from linear equations

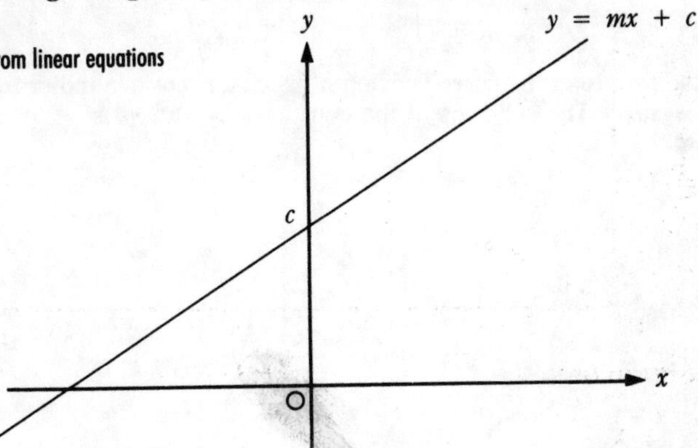

Quadratic equations

Graphs from quadratic equations of the form $y = ax^2 + bx + c$, where a, b and c are constants, are represented by a *U-shaped curve* (Fig G.7). We see that

Fig G.7

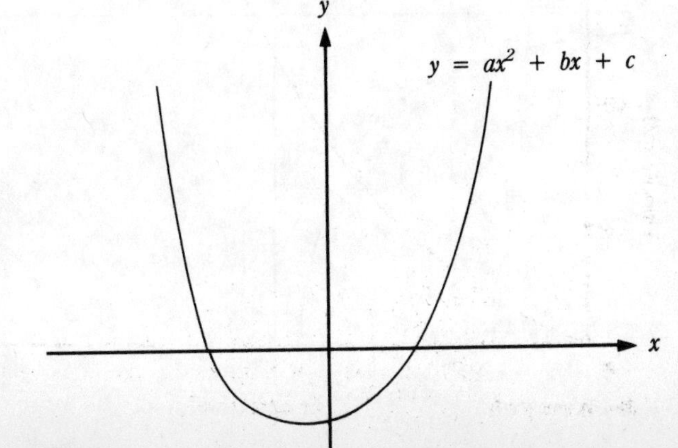

(i) a is positive

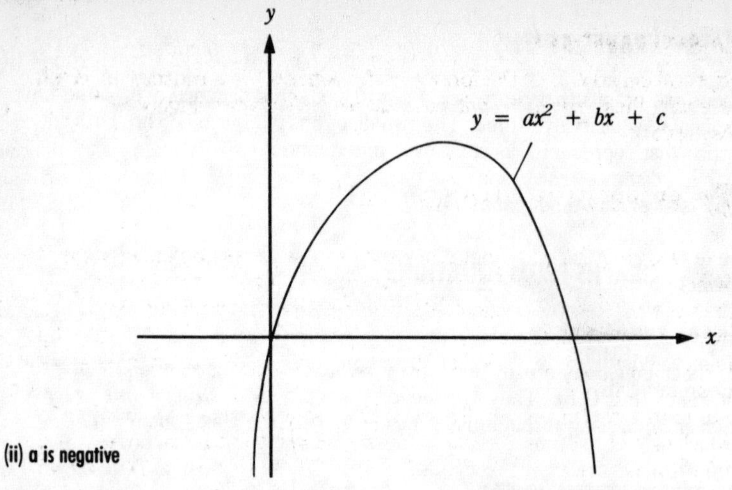

(ii) a is negative

the pattern of the curve is different when a is positive rather than when a is negative. The solutions of the equation $y = ax^2 + bx + c$ are the x axis intercepts.

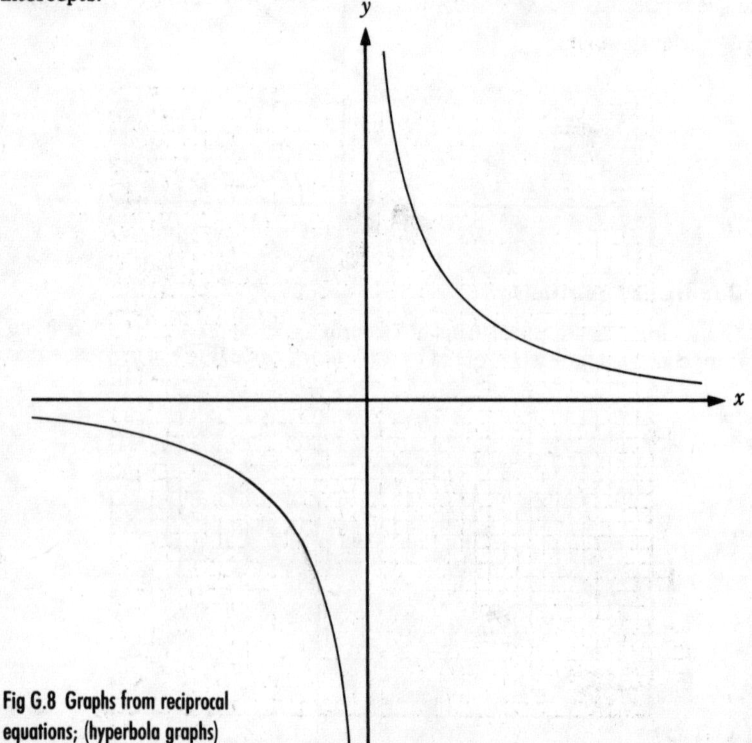

Fig G.8 Graphs from reciprocal equations; (hyperbola graphs)

Reciprocal equations

A reciprocal equation is of the form $y = A/x$ where A is a constant. A graph representing these equations will have the shape shown in Figure G.8.
◄ Asymptotic ►

GRAPHS OF INEQUALITIES

When drawing a graph of an **inequality** such as $y > 2x$ we need to indicate all the points on the grid where this inequality is true. There are a lot of points, but they will all fall on one side or the other of the straight line $y > 2x$.

If we were graphing the inequality $y \geqslant 2x$, the solution would include the points on the line $y = 2x$; whereas the solution of $y > 2x$ will not include the points on this line.

So to draw a region say $y < 3x$ we need to draw the line $y = 3x$, then to find which side of this line we want. One way of doing this is to choose some convenient point that is *not on the line*, say (0,2). (The *origin* (0,0) is always the simplest point to choose if you can, but of course in this case it is on the line itself, so you must choose another point). Substitute these values into the inequality and we get $2 < 0$; this is *not true*, so we shade-in the *other side* of the line rather than the side containing the point (0,2). See Figure G.9.

Fig G.9 Graphs of inequalities

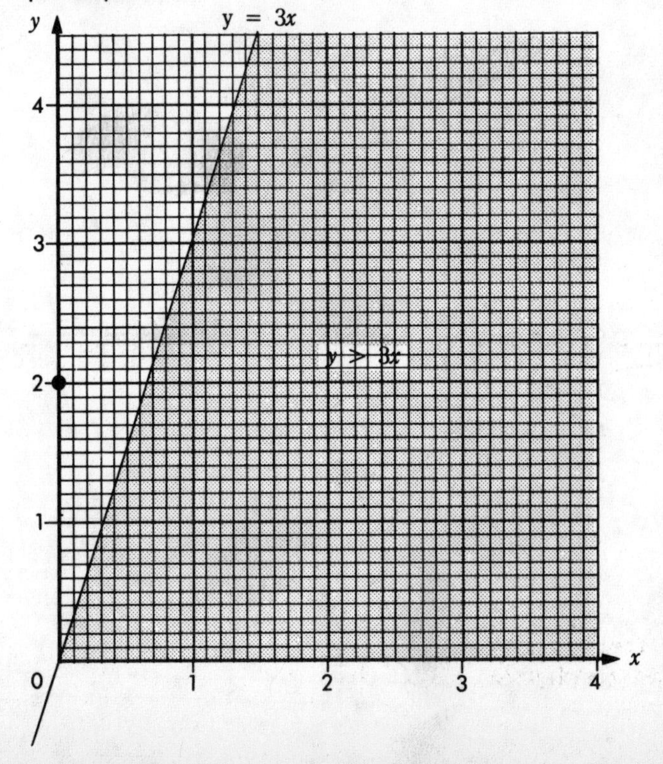

SKETCH GRAPHS

A sketch graph is a rough but reasonably accurate graph using as few points as possible to fix the main features of the graph. It is usually done by first recognising what type of graph it will be, whether linear, quadratic, reciprocal or something else. Then, remembering the main features of that type of graph, the necessary points that will fix the shape onto a grid are calculated and the graph sketched.

TRIGONOMETRICAL GRAPHS

Each of the trigonometrical functions has a recognisable graph that you ought to be familiar with.

Sine curve

The sine curve is the graph of $y = \sin x°$ (Fig G.10). Notice the main features are that it starts at the origin and the highest value of y is 1 with the lowest value -1. The curve has line symmetry about the line $x = 90°$ and $x = 270°$ although you would need to extend the graph both backwards and fowards to fully appreciate this.

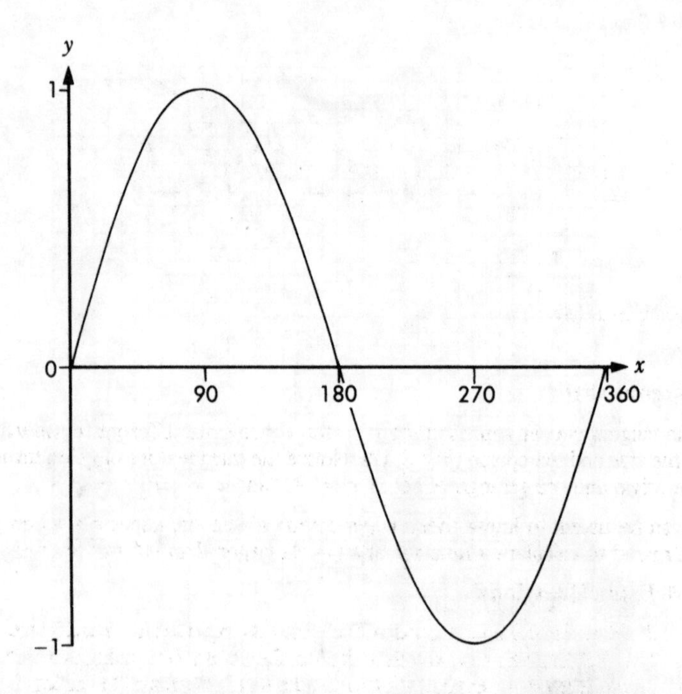

Fig G.10 Sine curve

Cosine curve

The cosine curve representing $y = \cos x°$ is the same as the sine curve, only it has been translated 90° (to the left) along the x axis (Fig G.11).

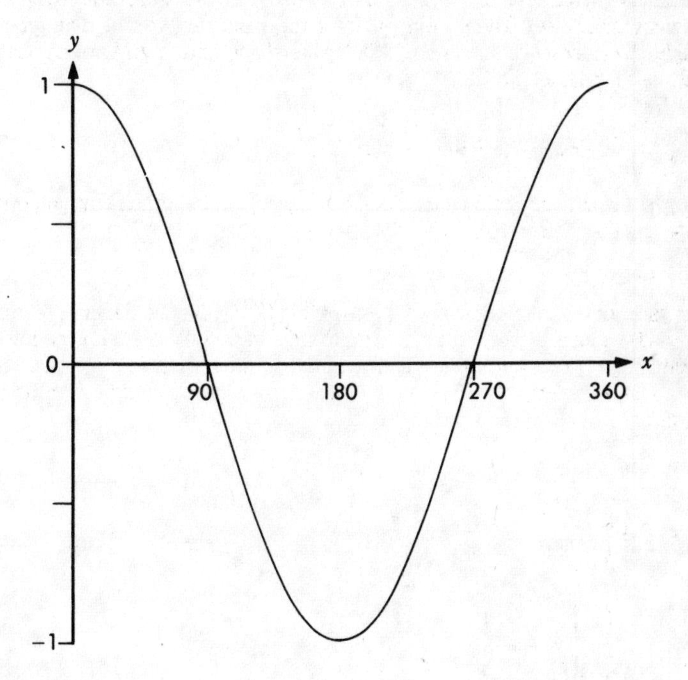

Fig G.11 Cosine curve

Tangent curve

The tangent curves representing $y = \tan x°$ are quite different to the waves of the sine and the cosine (Fig G.12). Notice the main feature of going through the origin and the strange effect at $x = 90°$ and $x = 270°$.

It can be useful to know these trigonometrical shapes, especially when you may need to calculate a function of an angle larger than 90° or vice-versa.

- **Exam Questions**

1 Water is sucked at a constant rate into the pipette shown in Figure G.13. On the axes provided, draw a sketch graph showing how the water level increases with time as the water is sucked to the level indicated on the drawing.

(NEA; H)

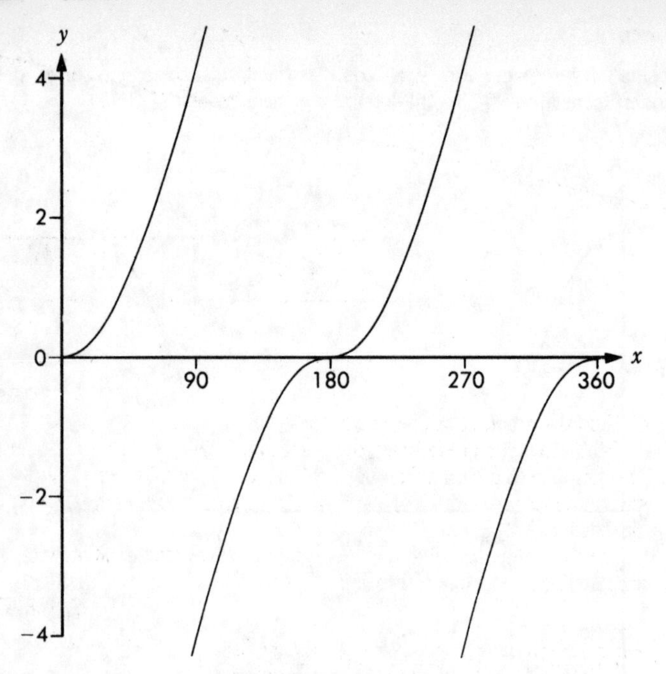

Fig G.12 Tangent curve

Fig G.13

2 The velocity v m/s of the ball in Figure G.14a) rolling down a slope is give by $v = 3t$, where t seconds is the time after the ball is released.
a) Complete the table in Figure G.14b) for values of v.
b) Draw the graph of v against t for values of t from 0 to 5.

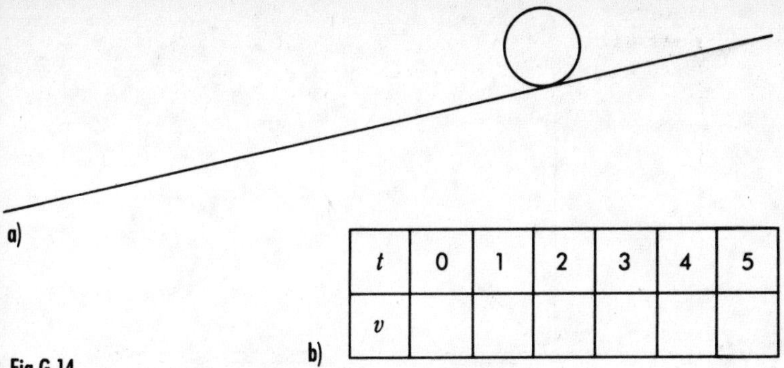

t	0	1	2	3	4	5
v						

Fig G.14

c) Find the gradient of this graph.
d) What does the gradient of this graph represent?

3 The graph in Figure G.15 shows the results of a pupil's experiment, where the total length (l) of a spring was measured when different masses were attached.

Use the graph drawn to find the equation linking the total length, l, with the mass, m, for this spring.

(NEA; H)

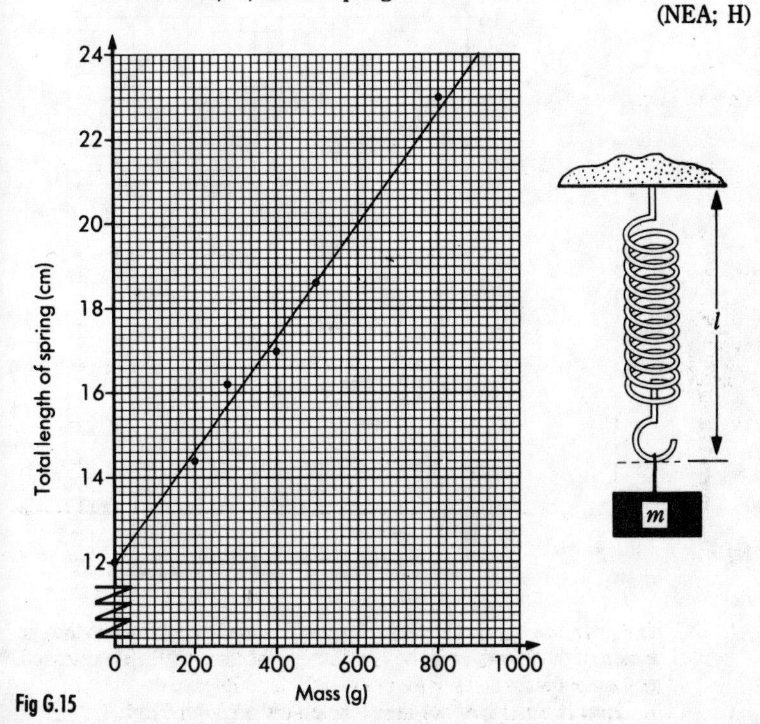

Fig G.15

Solutions

1 See Fig G.16.

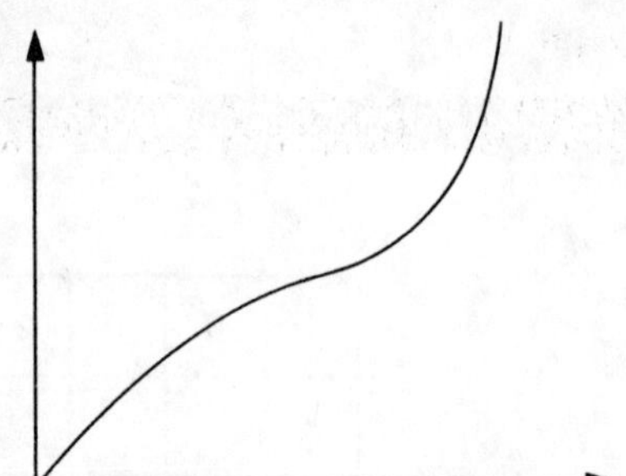

Fig G.16

2 a)

t	0	1	2	3	4	5
v	0	3	6	9	12	15

b) See Fig G.17.

c) Gradient = vertical distance ÷ horizontal distance
= 15 ÷ 5 = 3

d) The acceleration

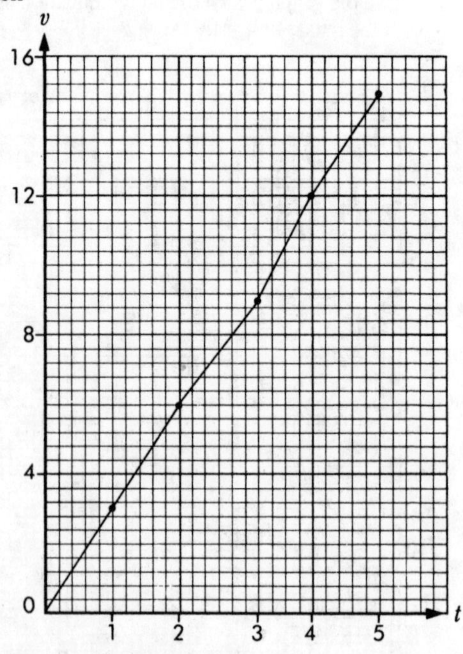

Fig G.17

3 The equation is linear; hence the form of $l = am + c$ where a is the gradient of the line and c is the l axis intercept.
The gradient is found by $12/900 = 0.013$
The l intercept is 12; hence the equation will be $l = 0.013m + 12$.

CONVERSION GRAPHS

◀ Conversion graph ▶

GROUPED DATA

◀ Data ▶

HEXAGON

A hexagon is a **polygon** with 6 sides.

HIGHEST COMMON FACTOR (HCF)

This is the highest of all the common **factors** of two or more whole numbers. For example, the common factors of 36 and 48 are 1, 2, 3, 4, 6 and 12. Since 12 is the highest, this is the highest common factor, or HCF for short.

HIRE PURCHASE

Usually called HP for short, hire purchase is sometimes called the 'never-never' or 'buy now, pay later'. It is a convenient way of spreading payment for goods over a period of some months or even years. It usually requires a **deposit** to be paid before the goods can be taken away and a promise (contract) to pay so much each week or month, as necessary. For example:

John bought his flute on HP. It originally cost £300, but the terms were £50 deposit and £15 a month for 24 months.

Note how the total paid will have been £$(50 + 15 \times 24)$ which is £410. The HP price is *never* less than the original cost price.

HISTOGRAM

A histogram is similar to a **bar chart**, but with the areas of the bars representing the frequency, and not the lengths of the bars. There should be no gaps between each bar in a histogram, as often happens with bar charts. The vertical axis is a **frequency density** and not frequency, so that the frequency is found by multiplying the width of the bar by the number on the frequency density axis.

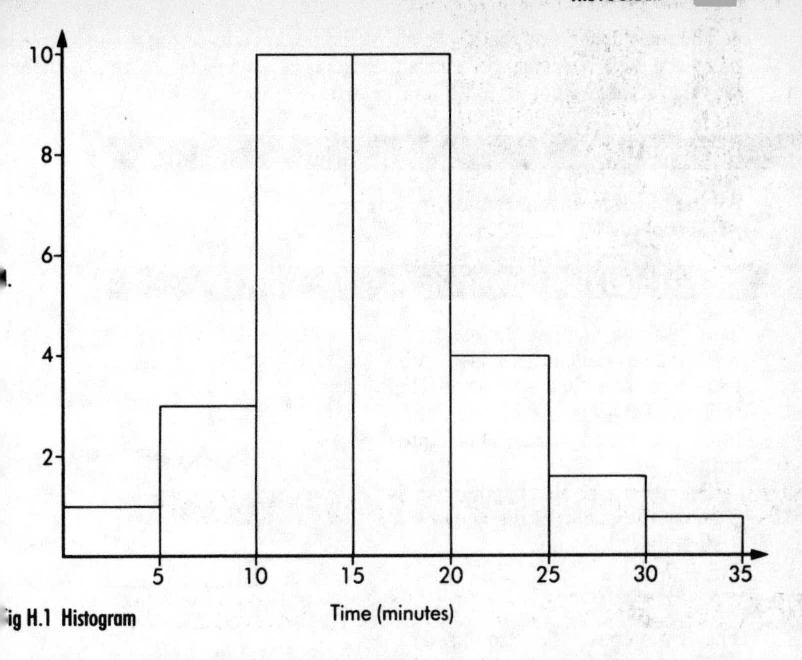

Fig H.1 Histogram

The histogram in Figure H.1 shows the waiting times that patients had at a doctor's surgery during one month. Notice how the waiting times between 10 and 20 minutes have been lumped together. Also notice that the smaller number of waits, between 0 and 5 minutes and 30 and 35 minutes, are such that we can still see them on the diagram without it looking ridiculous. From this histogram the groups are as in Figure H.2.

When drawing the *horizontal* scale we have to bear in mind the *type* of data that is being represented.

If the data is **continuous data**, rounded off to the nearest whole minute, then the vertical lines for the bars should start from 0 and 4.5; 9.5 and 19.5; etc., since these points truly represent the real range of each group.

However, if the data is **discrete**, then the vertical lines for the bars ought to be drawn at points that best represent the data, which could well be 0, 5, 10, 20, etc., in this case.

Waiting time	*f.d. × width*	*Frequency*
0–5	1 × 5	= 5
6–10	3 × 5	= 15
11–20	10 × 10	= 100
21–25	4 × 5	= 20
26–30	1.6 × 5	= 8
31–35	0.8 × 5	= 4

Fig H.2

The main concern is that there should be no doubt as to the group in which any piece of data will fall. So where the data is labelled 5, 10, 15 etc., it should be clear where values of 5, 10 and 15 actually are.

HYPERBOLA

A type of curve having two **asymptotes**.

◀ Graphs ▶

HYPOTENUSE

This is the longest side of a right-angled triangle; it is also to be found opposite the right angle (Fig H.3). Its length is often found by the theory of **Pythagoras**. This states that:

'the square of the hypotenuse is equal to the sum of the squares of the other two sides'.

ie $h^2 = a^2 + b^2$

$h = \sqrt{a^2 + b^2}$

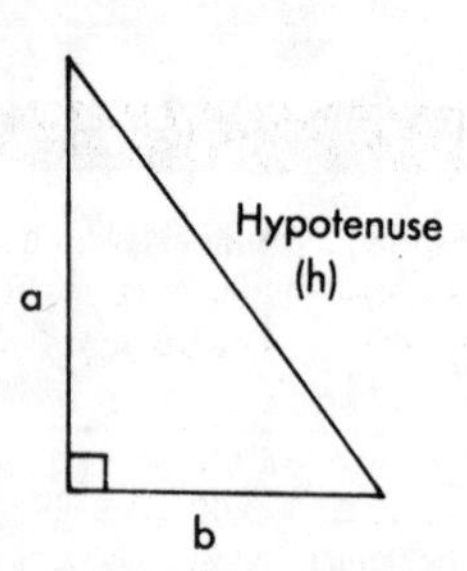

Fig H.3 Hypotenuse

The hypotenuse is also used in trigonometry where:

$\sin x$ = side opposite/hypotenuse
$\cos x$ = side adjacent/hypotenuse

HYPOTHESIS

The term *hypothesis* refers to some theory. In statistics we often wish to *test* hypothesis or theories to see whether we can be confident that their productions will hold true.

◀ Null hypothesis ▶

IDENTITY

In **geometry**, an identity is something which, when combined with anything else, leaves it unaltered. For example, an identity **transformation** is a transformation that leaves a shape where it is, e.g. rotation through 360°.

In terms of **algebra**, an identity is an expression which is true for *all* values of the variables in question.
For example, in the case of $x^2 - 6x + 9 \equiv (x - 3)(x - 3)$, the use of *three* horizontal lines indicates an identity. Whatever value of x you put in the left hand side of the expression, the expression will hold true on the right hand side for that value.
◀ **Unit matrix** ▶

IMAGE

An image is the result of some mathematical **function** or a **transformation**. For example, in the function defined as $f{:}x \rightarrow 5x$ then the image of x is $5x$, and so the image of, say, 3 is 5×3 which is 15.

Another example is a **reflection** of a shape in the y axis. Its image is the shape drawn after the reflection. The term 'image set' is sometimes used for the **range**.

IMPERIAL

Imperial is the word used to describe the weights and measures that belong (or belonged) to the official British series of weights and measures. The common units that you ought to be familiar with are:

12 inches = 1 foot
3 feet = 1 yard
16 ounces = 1 pound
8 pints = 1 gallon

There are others, but these are the ones that you do need to have some familiarity with. You ought also to be aware of the following **metric** equivalences:

2 pounds weight	is approximately	1 kilogram
3 feet	is approximately	1 metre
5 miles	is approximately	8 kilometres
1 gallon	is approximately	4.5 litres

INCOME TAX

Income tax is the type of tax that everyone who receives money for working, or from investments, has to pay to the government. This amount can change every time the government decides to change it (usually in the Budget). To calculate the amount of income tax you should pay, you need to know the *rate of tax* (a percentage) and your *personal* allowance.

Rate of tax

The rate of tax is expressed as a percentage, for example 25%. This means that you would pay 25% of your *taxable income* to the government as tax. This rate is sometimes expressed as a rate in the £. For example, the rate of 25% could be expressed as 25p in the £.

Personal allowances

Personal allowances are the amounts of money you may earn *before* you start to pay tax. They are different for single men and married men (at the moment), and for women in different situations.

Taxable income

You only pay *income tax* on taxable income which is found by subtracting your personal allowances from your actual annual income. If your personal allowances are *greater than* your annual income, then you pay no income tax.

Here is an example of income tax calculation:

Mr Coefield, who earns £19,700 a year has personal allowances totalling £4,300. What income tax does he pay in a year where the rate of tax is 25%?
The taxable income is 19,700 − 4,300 = 15,400
The income tax paid = 25% of £15,400 = £3,850.

INDEX

An index is the figure that is found at the top right hand corner of another number or algebraic expression to indicate a **power** (or **exponent**). The index tells us how many times that number is to be multiplied by itself. For example, $2^3 = 2 \times 2 \times 2$, the *three* being the index.

Standard index form

Standard index form is a convenient way of writing very large or very small numbers. It is always expressed in the terms of $A \times 10^N$ where A is a number between 1 and 10 and *N* is an integer (whole number). For example:

350 would be written as 3.5×10^2

413,200	would be written as	4.132×10^5
6,450.9	would be written as	6.4509×10^3

Notice how the index on the 10 tells you how many places *to the right* to move the decimal point.

If the number is less than 1 to start with, then the index on the 10 will be negative. Here the index will tell you how many places *to the left* to move the decimal point. For example:

0.045	would be written as	4.5×10^{-2}
0.0008	would be written as	8.0×10^{-4}

INDICES

Indices is the plural of index, and is the word used to describe the *set of numbers* written in this index notation, e.g. x^6.

Rules of indices

If the *base number* is the same in each case then:

- When you multiply indices you simply need to *add* their powers:
  i.e. $a^x \times a^y = a^{x+y}$
  e.g. $y^2 \times y^3 = y^{2+3} = y^5$
- When you divide indices you simply need to *subtract* their powers:
  i.e. $a^x \div a^y = a^{x-y}$
  e.g. $y^5 \div y^2 = y^{5-2} = y^3$
- You need to know that for any number x then $x^1 = x$ and $x^0 = 1$
- We use *negative* indices to show numbers of a fractional form.
  For example:
  $x^{-3} = 1/x^3$ and $3^{-2} = 1/3^2$.
- Fractional indices can also be used to indicate *roots*. For example:
  $16^{1/2}$ will be $\sqrt{16}$ which is 4 and -4.
  Also $x^{1/3}$ will indicate $\sqrt[3]{x}$, that is the cubed root of x.
  Hence the power $1/n$ will indicate the nth root.
- Note that a power such as 2/3 will indicate the square of the cubed root, or the cubed root of the square. For example:
  $27^{2/3} = \sqrt[3]{(27^2)} = \sqrt[3]{729} = 9$
  or $(\sqrt[3]{27})^2 = 3^2 = 9$

- **Exam Question**
  Find the value of x in each of the following equations:

  a) $2^x = 8$
  b) $2^x = 1/8$
  c) $8^x = 1/2$

(SEG; H)

- **Solution**
  a) $x = 3$
  b) $x = -3$
  c) $x = -1/3$

INEQUALITIES

Inequalities are signs used to indicate how two expressions might be different in terms of their relative sizes:

$>$ means greater than, e.g. $5 > 2$
$<$ means less than, e.g. $3 < 6$
$\geqslant$ means greater than or equal to
$\leqslant$ means less than or equal to

INEQUATIONS

An inequation is an equation with an ***inequality sign*** in it; but the term **inequality** is often also used to describe an inequation. They are solved in much the same way as a normal equation is solved, except for when you wish to multiply or divide both sides by a ***negative amount.*** In this case you then have to ***reverse*** the inequality sign, for example $3 > 1$ would become $-3 < -1$. For example:

Find the range of values for which $2x - 7 > 3(5 - x)$
Using normal equation techniques we can calculate that
$2x - 7 > 15 - 3x$
to give $5x > 22$ so $x > 22/5$
hence the solution is $x > 4.4$

GRAPHING INEQUATIONS

Graphs of inequalities lead us into *solution sets*, which is when we need to find a solution to a *set* of inequations. In this case we would shade all unwanted regions, leaving ourselves with the possible solution set ***unshaded.*** For example:

Illustrate the set of points that satisfy the inequations: $x > 0$, $y > 0$, $x + y < 6$, $x + 3y > 6$.
The $x > 0$ and the $y > 0$ indicate to us that we only need the region where both x and y are positive. Hence shading out the other two regions leaves us with the possible points we are looking for as in Figure I.1.

Always be aware of the inequality sign and remember to notice whether it is $>$ or $\geqslant$ etc. In the latter case you do need to ***include the points on the line*** in your solution set.

This is a difficult part of graph work and an area where a lot of mistakes are so easily made, especially in finding out the various regions and correctly shading them. Be very careful to check your inequations and to make certain that you are shading the correct region, as required by the situation.

◀ Graphs ▶

INFINITE

This is a number that is far too big to be counted, or indeed, is uncountable; for example, the number of stars in the sky, or the number of grains of sand in

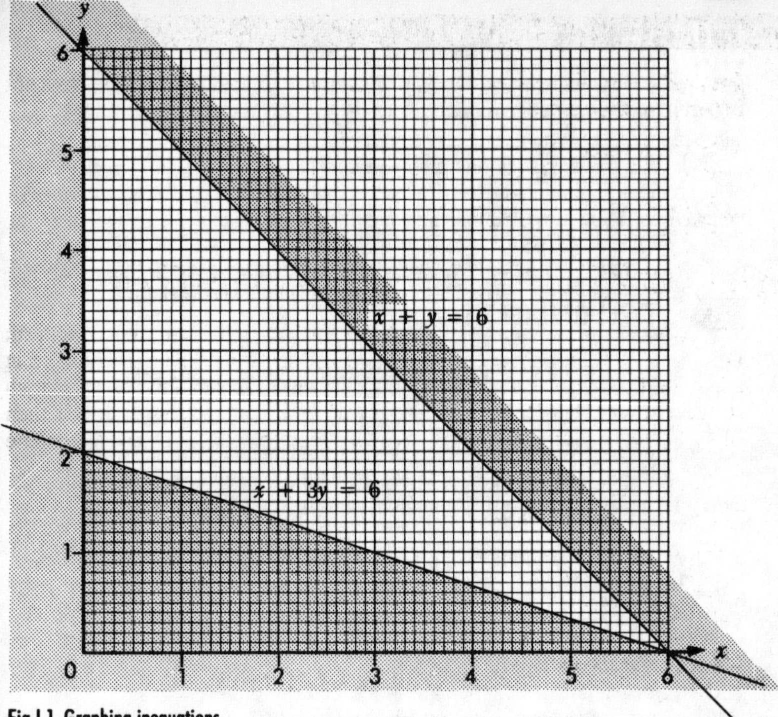

Fig I.1 Graphing inequations

the world, or the number of times you can take 0 away from 1. This last example indicates just why 1/0 is infinity and why, when you try it on your calculator, you get E for error, since the calculator cannot do it.

◀ Sets ▶

INTEGERS

An integer is a whole number like 1,2,3,4 etc. It can either be positive or negative as with −1, −2, −3 etc.

INTERCEPT

An intercept is where a line crosses an axis; an intercept on the x axis is called an *x axis intercept*, and an intercept on the y axis a *y axis intercept*.

In the general equations of lines, the y axis intercept is given by the *constant* in the equation:

- for a *linear equation* of $y = mx + c$, then c is the y axis intercept
- for the *quadratic equation* of $y = ax^2 + bx + c$, again c is the y axis intercept.

These points can be most useful when sketching graphs.

INTEREST

Interest is what we call the amount of money that someone will give you for letting them borrow your money, or what you pay for borrowing money from someone else. So you can receive interest and you can *pay* interest.

Banks and building societies give you interest if you let them borrow some of your money. For example, if a bank pays 6.5% interest *per annum* (per year), then if you leave £30 in their bank for the one year, they will pay you £30 × $^{6.5}/_{100}$ = £1.95.

TYPES OF INTEREST

There are two types of interest, simple and compound interest.

Simple interest

Simple interest (SI) is calculated on the basis of having a *principal amount*, say P, in the bank, for a number of years T, with a rate of interest R. There is then a formula to work out the amount of simple interest earned:
SI = PRT/100
In other words, you multiply the principal by the rate by the time and then divide by 100. For example:

Joseph had £14.50 in an account that paid simple interest at a rate of 8%. Calculate how much interest would be paid to Joseph if he kept the money in this account for 4 years.
The principal is £14.50, the rate is 8% and the time is 4 years.
Hence using the formula SI = PRT/100 = 14.5 × 8 × 4/100 which is £4.64

Compound interest

Compound interest (CI) is the type of interest most likely to be paid to you by banks and building societies. It is based on the idea of giving you the simple interest after 1 year and then adding this onto your principal amount (sometimes this interest is calculated and added every 6 months). The money then grows more quickly than it would with just simple interest. For example:

Helen paid £50 into a bank that paid her 5% interest every 6 months, adding this to the principal every 6 months. Calculate how much she has in the account after 2 years.

- After 6 months she earns interest of £50 × 5/100 which is £2.50; hence she will have £52.50
- After 12 months she earns interest of £52.50 × 5/100 which is £2.63; hence she will have £55.13
- After 18 months she earns interest of £55.13 × 5/100 which is £2.76; hence she will have £57.89
- After 2 years she earns interest of £57.89 × 5/100 which is £2.89; hence she will have £60.78

Note that there is a formula for working out compound interest:
$$CI = P \times (1 + {}^{R}/_{100})^{N} - P$$

where P = Principal amount, R = Rate of interest, N = Number of times to be applied.
The formula for the *final amount* earned in compound interest is:
Total Interest = $P \times (1 + R/100)^N$

Notice how the formula is simpler if you want to find the whole amount. But it must be used with caution, because if the interest is actually paid into an account each year, or 6 months, etc., then rounding off will be taking place which will eventually give a slightly different answer to the formula. In an exam situation, either method of gaining the answer to compound interest will be acceptable; but you ought to be aware that there is a difference. Try this out on the example above if you wish to check for yourself.

- **Exam Question**

  In a savings account, compound interest is paid at 8.5% per year. Jim starts with £200 in his savings account. Calculate:

  a) how much interest will be paid in the first year,
  b) the total in his account after one year,
  c) how much interest will be paid in the second year,
  d) the total in his account after two years.

  (NEA; I)

- **Solution**

  a) $8.5 \times 200 \div 100 =$ £17
  b) $200 + 17 =$ £217
  c) $8.5 \times 217 \div 100 =$ £18.45
  d) $217 + 18.45 =$ £235.45

INTERQUARTILE RANGE

This is the difference between the upper and lower **quartiles** on any cumulative frequency curve. It should be expressed simply as the number difference on the horizontal axis. For example:

In Figure I.2, the upper quartile is 20 and the lower quartile is 10.5, so the interquartile range is $20 - 10.5$, which is 9.5.

The use of interquartile ranges is to define a range of marks and to see how a frequency distribution is spread out.

Semi-interquartile range

The semi-interquartile range is exactly what it says – half of the interquartile range. So, divide the interquartile range by two.

- **Exam Question**

  Figure I.3 illustrates the cumulative frequency distributions for the heights of two different types of mature (fully grown) corn, type A and type B.

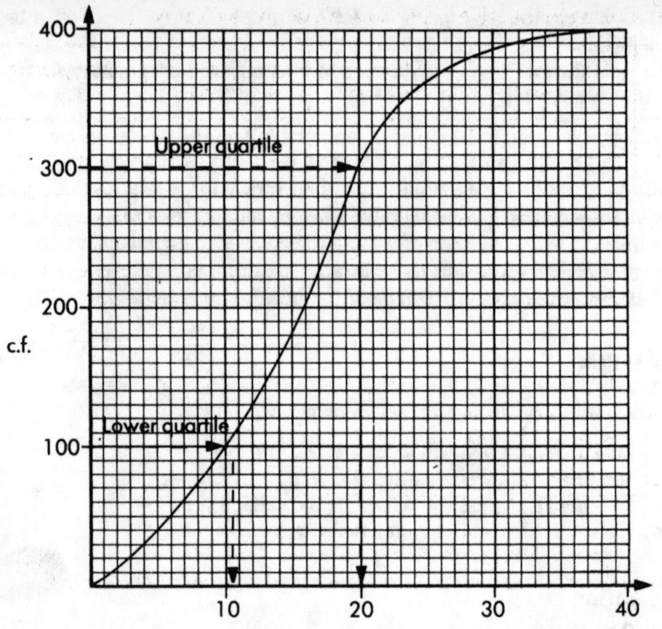

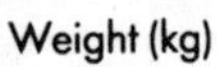

Fig I.2 Interquartile range

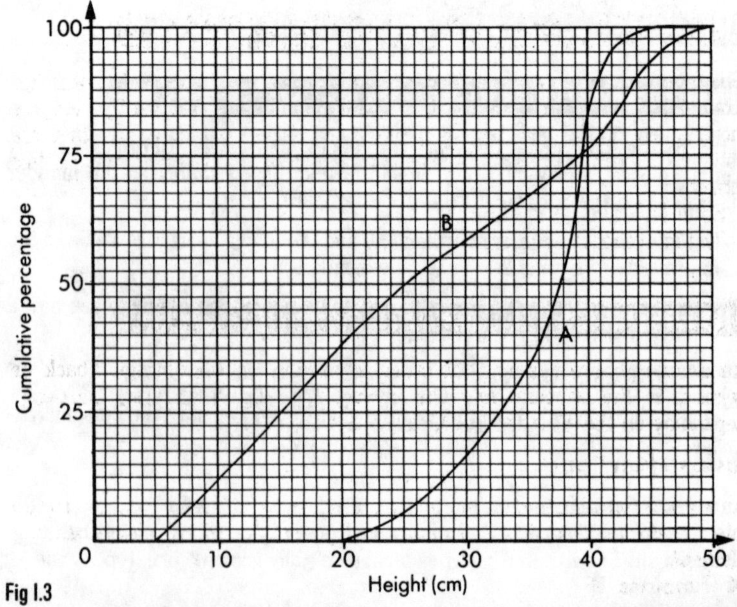

Fig I.3

a) Complete the table below for the two types of corn.

	Median height (cm)	*Upper quartile (cm)*	*Lower quartile (cm)*	*Interquartile range*
Type A				
Type B				

b) Comment briefly on the distributions of the height of the two types of corn.

c) A mature corn plant is measured and found to have a height of 28 cm. State, with a reason, which of the two types of corn you think it is.

(NEA; H)

■ **Solution**

a)

	Median	*U.Q.*	*L.Q.*	*I.Q.R.*
A	37.5	39.5	33	6.5
B	25	39.5	15	24.5

b) Type A will all grow to a similar height. Type B shows a wide spread of heights with little reliability in the growth.

c) Type B is the more likely because only a small proportion of type A corn would be this small.

INTERSECTION

The intersection of two sets is the elements that are common to both. For example, if A is the set {2,4,6,8} and B is the set {1,2,3,4,5} then the intersection of the two sets is {2,4}, these being the elements that are common to both sets A and B. The usual notation for intersection is ∩. For example:

State {a,b,c,d,e,f} ∩ {b,a,c,k}.
The intersection is {a,b,c}.

INVERSE

An inverse generally means 'the other way round' or 'what brings it back' or 'what it is the opposite to'. The inverse has slightly different meanings depending on the situation in which the term is used.

Inverse functions

An inverse function is the **function** that brings elements back from the **range** into the **domain**, so that each image will come back to its original starting element.

◀ Functions ▶

Inverse matrices

This will depend whether we want the *additive* inverse or the *multiplicative* inverse. In most cases what is meant by an inverse of a matrix is the multiplicative one, but both types could be meant.

- Additive inverse: the additive inverse of the matrix M is the one we need to add to M so that they give the **null matrix** of the same order. For example, the additive inverse of $\begin{pmatrix} 4 & 5 \\ 3 & 2 \end{pmatrix}$ is $\begin{pmatrix} -4 & -5 \\ -3 & -2 \end{pmatrix}$
- Multiplicative inverse: the multiplicative inverse of the matrix M is the one which, when multiplied to M, gives the matrix $\begin{pmatrix} 1 & 0 \\ 0 & 1 \end{pmatrix}$

  The multiplicative inverse is found by the following formula:

  for the matrix $\begin{pmatrix} a & b \\ c & d \end{pmatrix}$ it is $\frac{1}{(ad-bc)} \begin{pmatrix} d & -b \\ -c & a \end{pmatrix}$

  If this formula is required from you in the examination, it will almost certainly be on the *formula sheet* issued for the examination.

Inverse transformation

An inverse of a **transformation** T is that transformation T′ which, when combined with T, would leave any shape where it originally was.

- Reflection: **Reflections** have what are called **self-inverses** since their inverses are themselves. For example, for a reflection in the line $y = x$ . . . the *inverse* will be the same reflection in the same line.

Other transformations have simple-to-find inverses, for example:

- Rotations: **Rotations** around a point through an angle x. Here the inverse will be another rotation around the same point, but through the angle $-x$.
- Enlargement: **Enlargements** of scale factor K through some point. Here the inverse will be another enlargement through the same point, but of scale factor 1/K.
- Translation: **Translation** of a vector $\begin{pmatrix} a \\ b \end{pmatrix}$.

  Here, the inverse will be another translation with a vector $\begin{pmatrix} -a \\ -b \end{pmatrix}$.

Most transformations within a GCSE syllabus can be represented by a **matrix**, and the *inverse transformation* will be represented by the *multiplicative inverse matrix*.

INVESTIGATIONS

These are mathematical enquiries where you need to search through selected data in order to find a pattern or a solution to some defined problem.

Investigations are best examined through **coursework**. However, there will be an element of investigation within examination questions, particularly where you need to examine some number sequence.

In explaining any solution to your investigation, you must make all your reasons clear, stating ***how*** you found any relationships. For example:

Some numbers like 4 and 9 have exactly three factors and no more. Find the next three numbers like these to have exactly three factors.

After a short search, you should find that 25 is the next number to have exactly three factors. Note that we have square numbers but only the squares of **prime numbers**; so the next few such numbers will be 7×7 and 11×11. Hence the next few numbers will have been 25, 49 and 121, but you should have ***stated the reasons*** why these are the next three numbers.

■ Exam Questions

The questions below refer to the numbers in the grid.

0	3	6	9	12	....
4	7	10	13	16	....
8	11	14	17	20	....
12	15	18	21	24	....
16	19	22	25	28	....
.	.	.	.	.	
.	.	.	.	.	
.	.	.	.	.	
.	.	.	.	.	

The diagonal difference for any 2×2 square on the grid is defined as $qs - pr$, where p, q, r and s are numbers in the square as shown below.

p	q
s	r

Squares are identified by the number in the top left-hand corner. This square is called a 'p' square.

a) Find the diagonal difference for the '14' square.

b) Investigate the diagonal difference for **two** other squares on the grid. Write down your results and any observations that you can make.

(NEA; H)

■ Solution

a) The 14 square is

14	17
18	21

The diagonal difference will be $(18 \times 17) - (14 \times 21) = 12$

b) The other two you did will also give you a diagonal difference of 12.

IRRATIONAL

An irrational number is one that *cannot* be expressed as a **vulgar fraction**, i.e. as a/b, where both a and b are integers (whole numbers). It is also true to say that an irrational number *cannot* be expressed as a **terminating** or **recurring** decimal.

The most common examples of irrational numbers are π and $\sqrt{2}$.

ISOSCELES

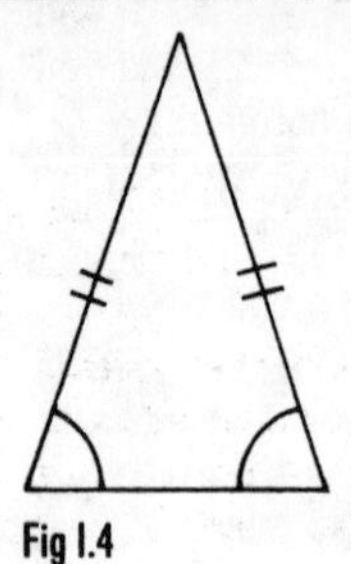

Fig I.4

An isosceles triangle has two of its sides the same length and two angles the same, as in Figure I.4. The line that bisects the angle included between both sides of the same length is a **line of symmetry** and is **perpendicular** to the line facing the angle. This is useful to know when asked to find angle sizes or lengths, since you can then use **trigonometry**. For example:

Find the size of angle x in Figure I.5.

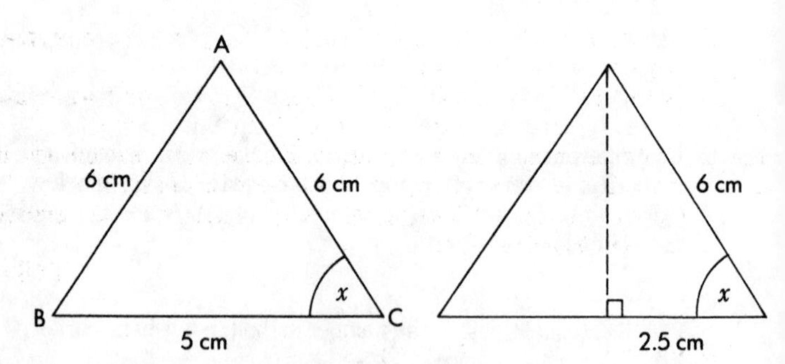

Fig I.5

First, drop a perpendicular down from A to BC. This will divide the triangle in half, to give us the right-angled triangle as shown. Now use trigonometry to calculate the size of angle x as 65.4°.

ITERATION

An iteration is when some generating term, U_n, is used to keep on generating terms until a certain situation is satisfied. For example, a solution to the equation $x^2 - 2x - 3 = 0$ can be found by re-writing the equation in the form: $x^2 = 3 + 2x$, then in the form:

$$x = \frac{3}{x} + 2$$

Suppose we assume a starting solution X_1 to this equation as being $X_1 = 2$.

We now find the value this starting solution makes the right hand side of the equation. We will find that:

$X_2 = 3/x + 2 = 3/2 + 2 = 3.5$

We then use this value of $X_2 = 3.5$ as a better solution in the equation. We now get $X_3 = 3/3.5 + 2 = 2.857 \ldots$

By continuing the process we find that:

$X_4 = 3.05$
$X_5 = 2.98$
$X_6 = 3.01$
$X_7 = 3.00$
$X_8 = 3.00$

The process was continued *until* the value to 2 decimal places was the same two times in succession. The actual calculator value was used each time in the iteration. Hence the solution here is $x = 3$, which can be shown to be correct by substituting it into the original equation.

This type of question in an examination will almost certainly start by leading you to the first starting solution. The most common error for candidates to make is to *round off* in the middle of the iteration, instead of keeping the correct values in the calculator. By all means show rounded off values in your method of solution, but keep as accurate a value as possible in your calculator.

■ Exam Questions

a) The iterative formula $U_{n+1} = \frac{1}{10}(12 - U_n^2)$ is used to generate a sequence of numbers, given the value of U_1.
Given that $U_1 = 1$, find U_2, U_3, U_4 and U_5, giving your answers to U_4 and U_5 correct to four places of decimals.

b) Using the formula for solving quadratic equations, solve the equation $x^2 + 10x - 12 = 0$ correct to two places of decimals.
Comment on the relationship between your solution to this equation and your answers to part a).

(NEA; H)

■ Solution

a) $U_2 = 0.1\,(12 - 1^2) = 1.1$
$U_3 = 0.1\,(12 - 1.1^2) = 1.079$
$U_4 = 0.1\,(12 - 1.079^2) = 1.0836$
$U_5 = 0.1\,(12 - U_4^2) = 1.0826$

b) Use $x = \dfrac{-b \pm \sqrt{(b^2 - 4ac)}}{2a}$

$= \dfrac{-10 \pm \sqrt{(100 + 48)}}{2} = -11.08$ and 1.08

The positive solution of 1.08 is the same as the solution to the iteration.

KITE

A kite, recognisable as the shape of a kite, has four sides, as in Figure K1, with the top two sides the same length and the bottom two sides the same length.

A kite will have one **line of symmetry** bisecting the angles included between the sides of the same length.

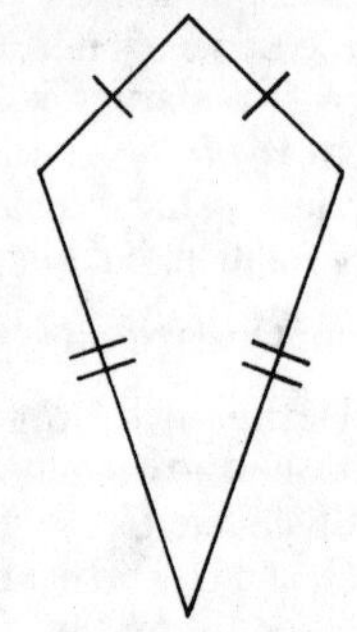

Fig K.1 Kite

LENGTH

Length is a distance from one particular point to another. The length of a line is the distance from one *end* to the other. Examination questions often ask you to find a length; you usually need to use either **trigonometry** or the theorem of **Pythagoras.**

LINEAR

Linear has to do with begin straight; so a line that is linear is straight. The general equation of a linear line is $y = mx + c$ where m is the gradient of the straight line and c is the y axis intercept. This **equation** is called a *linear equation.*

◀ Equation, Graph ▶

LINE OF BEST FIT

A line of best fit is the line that is drawn on a **scatter diagram** that best appears to represent the trend of the situation being graphed. This line is normally drawn as a *straight line*, but not necessarily so; it could be *curved* if

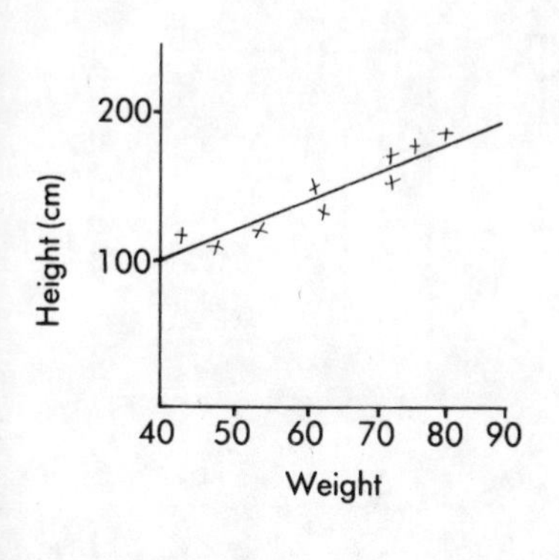

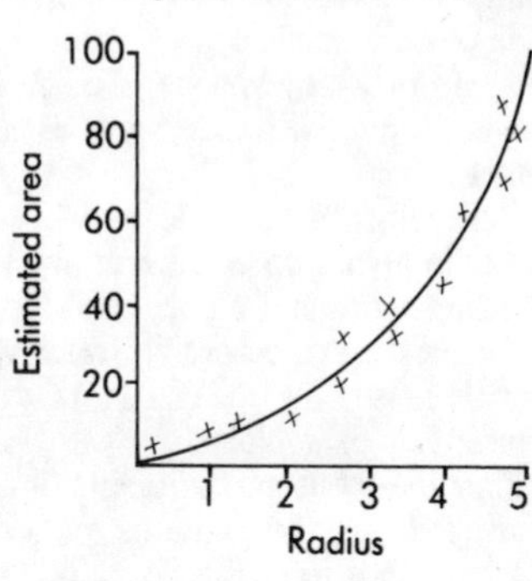

Fig L.1 Line of best fit

that is what the trend is. However, the line of best fit is always the simplest line that could be drawn.

Look at the two scatter diagrams in Figure L.1. The line of best fit has been drawn in both cases. Note how one is a straight line and the other a curve.

Strictly speaking, the line of best fit is that line which minimizes the sum of squared deviations from the line. In later statistics courses you can use a formula to find the line of best fit.

LINE OF SYMMETRY

If you can fold a shape over so that one half fits exactly on top of the other half, then the line over which you have folded is called a line of symmetry. The examples in Figure L.2 illustrate the lines of symmetry in various shapes as dotted lines. The square has four lines of symmetry, the rectangle two, the isosceles triangle just one, as has the pentagon next to it, while the circle has thousands and thousands of lines of symmetry: there are too many for us to count (we call this an infinite number).

Often, of course, you cannot fold over a shape that you are looking at, so you either have to imagine it being folded or trace it on tracing paper and then fold it. In most examinations you would be allowed to trace the shape and fold it over to find lines of symmetry.

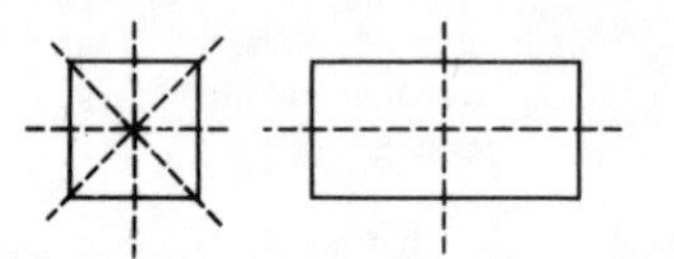
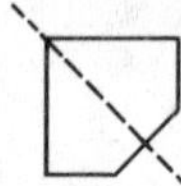
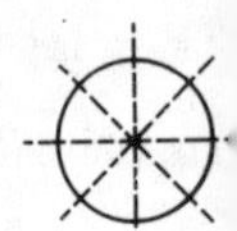

Fig L.2 Lines of symmetry

LOCUS

A locus is the collection of all possible points relating to some rule. For example, the locus of all points 2 cm away from a dot will be a *circle* of radius 2 cm with that dot as its centre.

Generally, to find a locus we can fix some points on a drawing and go on fixing more and more points until we see a pattern emerging which we can then say is the *locus* of the points. For example:

Find the locus of the points 1 cm away from the square, of side 4 cm, shown in Figure L.3. The locus is the dotted line and was found by fixing in all those points

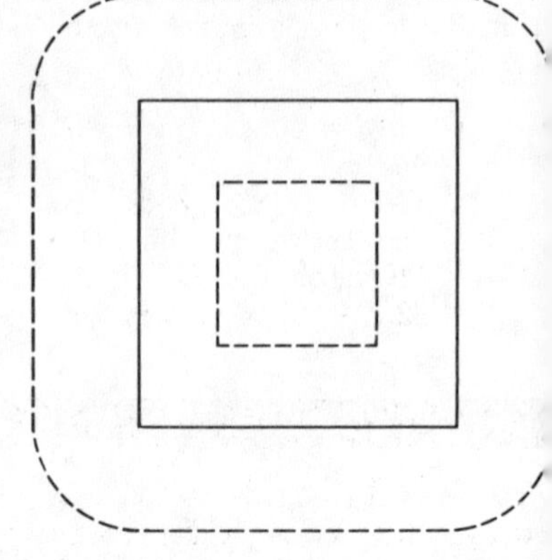

Fig L.3 Locus

1 cm away from the edges of the square. Around each *vertex* there is a quarter-circle, meeting lines which are parallel to, and equal in length with, each side. Then *inside* the square, the locus continues as another square of side 2 cm.

The most common mistake to be made on locus questions is for students to think they have a pattern before trying it out fully. As a result, only *part* of a pattern is often found, which of course is a wrong locus and so few marks would be gained.

■ **Exam Question**

Figure L.4 represents a rectangular yard adjacent to a factory, which is patrolled by a guard dog, D, tethered by a chain 20m long attached at the mid-point, M, of the factory wall, SR.

a) Assuming the dog keeps the chain taut, sketch on the diagram the locus of the dog's path as it moves from one side of the yard to the other.
b) Calculate the distance PM.
c) An intruder climbs into the yard at P. Mark on the diagram the point N at which the dog is closest to the intruder. Calculate the distance PN.

(NEA; I)

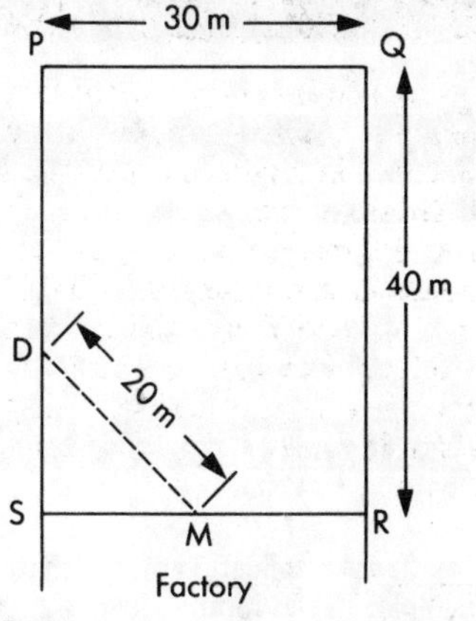

Fig L.4

■ **Solution**

a) See Figure L.5.
b) By Pythagoras,
$PM^2 = 40^2 + 15^2 = 1825$
so
$PM = \sqrt{1825} = 42.7$ m.
c) See N on the diagram, then PN will be
$42.7 - 20 = 22.7$ m.

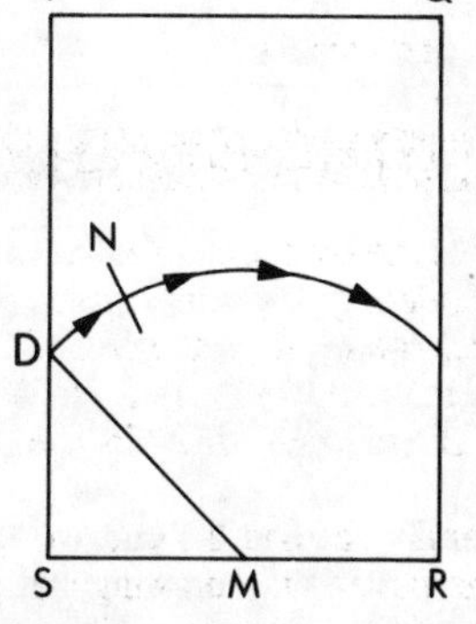

Fig L.5

LOWEST COMMON MULTIPLE

The Lowest Common Multiple (LCM) of a set of integers is the smallest integer that can be divided exactly by each of the integers in the set. For example, the lowest common multiple of 4, 6 and 8 is 24, since 24 is the smallest integer that each of the integers 4, 6 and 8 will divide into exactly.

MAPPING

◀ Function, Image ▶

MATRIX/MATRICES

A matrix is a collection of information put in tabular form in a *specific order*. For example, in the football league table illustrated below, the numbers form the matrix. They are in a specific order, as indicated by the labelling at the top and up the side:

LEAGUE TABLE

Team	*played*	*won*	*drawn*	*lost*	*for*	*against*	*points*
Sheff. Wed	40	30	6	4	86	25	96
Liverpool	40	30	4	6	81	28	94
Arsenal	40	24	8	8	65	35	80
Norwich	40	22	5	13	74	38	71
Coventry	40	18	7	15	75	41	63

ORDER OF A MATRIX

The **order** of a matrix is given by an ordered pair of numbers $a \times b$ where a is the number of *rows* in the matrix and b is the number of *columns*. So the football table shown above has an order of 5 by 7. Note that it is the *number* of numbers that make up the order.

MATRIX ARITHMETIC

Addition and subtraction

Since the information in the matrix is in a specific order we can *add* or *subtract* matrices as long as it is sensible to do so. In other words, we can add or subtract only when the information in each matrix is about the same thing and when the matrices are of the same order. We then add or subtract the corresponding numbers in each matrix, as with these two matrices of order 2×3.

For example: $\begin{pmatrix} 2 & 5 & 3 \\ 4 & 1 & 6 \end{pmatrix} + \begin{pmatrix} 1 & 7 & 2 \\ 2 & 8 & 3 \end{pmatrix} = \begin{pmatrix} 3 & 12 & 5 \\ 6 & 9 & 9 \end{pmatrix}$

Multiplication

Two matrices can only be *multiplied* together if the *second matrix has the same number of rows* as the *first matrix has columns.*

An (x by y) matrix multiplied by a (y by w) matrix, will have the order of (x by w).

Each *row* of the first matrix is combined with each *column* of the second matrix by *multiplying* the first row number by the first column number. Then the second row number is multiplied by the second column number, and so on until you reach the end of the row (and hence the end of the column also). You then *add together* all the products you have just made from that row and column. This *sum* becomes a new entry in the product matrix and in the position of the same row, same column as has just been combined. Here are two examples:

- $(3 \;\; 2 \;\; 6) \begin{pmatrix} 5 \\ 4 \\ 1 \end{pmatrix} = 3 \times 5 + 2 \times 4 + 6 \times 1 = (29)$

- $\begin{pmatrix} 3 & 2 & 6 \\ 2 & 1 & 0 \end{pmatrix} \begin{pmatrix} 5 & 2 \\ 4 & 1 \\ 1 & 3 \end{pmatrix} = \begin{pmatrix} 29 & 26 \\ 14 & 5 \end{pmatrix}$

- **Exam Question**

  Trackpaks of various sorts are sold for model railways.
  Trackpak *C* contains 5 Straights, 2 Points and 4 Curves.
  Trackpak *D* contains 2 Straights, 1 Points and 8 Curves.

  a) The contents of Trackpak C may be represented by the matrix

  $$C = \begin{pmatrix} 5 \\ 2 \\ 4 \end{pmatrix}$$

  Give a similar matrix **D** for the contents of Trackpak *D*.

  b) i) Evaluate the expression 3**C** + 2**D**.

  ii) Explain clearly what your answer to b)i) represents.

  c) The cost of the components is as follows:
  Straights 95p each, Points £2.15 each, Curves £1 each.

i) Evaluate the product $(0.95\ 2.15\ 1.00)\begin{pmatrix}5 & 2\\ 2 & 1\\ 4 & 8\end{pmatrix}$

ii) Explain clearly what your answer to c) i) represents.

(MEG; H)

■ **Solution**

a) $C = \begin{pmatrix}2\\ 1\\ 8\end{pmatrix}$

b) i) $\begin{pmatrix}19\\ 8\\ 28\end{pmatrix}$

ii) The number of straights, points and curves in 3 trackpak Cs and 2 trackpak Ds.

c) i) (13.05 12.05)

ii) The cost of a trackpak C and D respectively.

ROUTE MATRICES

One-stage route matrices

Given a network of paths, as in Figure M.1a), we can describe this by using a matrix to tell us how many direct one-stage routes there are from one point to another (e.g. A to B without going through any other points).

Hence the matrix that will describe the network is shown in Figure M.1b), and reads from left to right. Note that from C to C there are two routes, since we can go either way along the path shown.

Two-stage route matrices

A **two-stage route** is one that combines *two* one-stage routes to get from one point to another. For example:

Find the two-stage route matrix for the network in Figure M.2.

Fig M.1 One-stage route matrices

a)

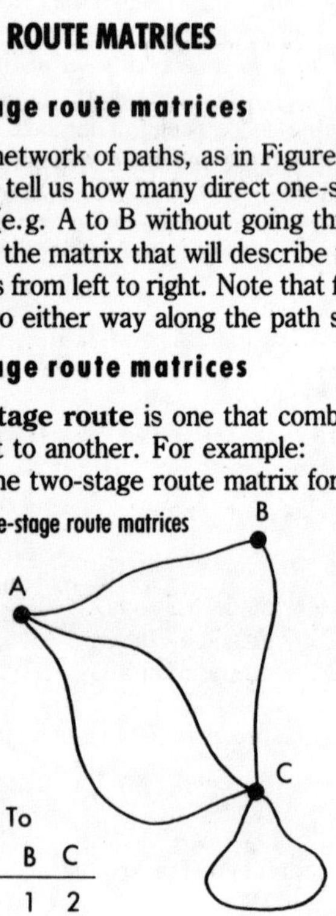

A B C

b)

From \ To	A	B	C
A	0	1	2
B	1	0	1
C	2	1	2

Fig M.2 Two-stage route matrices

(a

A B C

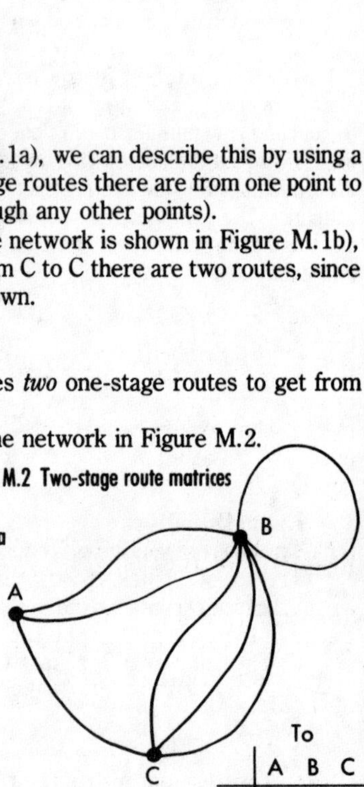

b)

From \ To	A	B	C
A	5	7	6
B	7	17	8
C	6	8	10

This is found by considering each pair of points and finding the number of two one-stage routes between them:

- Hence A to A has 5 different routes:
  1) A to B, top road, and back using the top road to A
  2) A to B, top road, and back using the bottom road to A
  3) A to B, bottom road, and back the same way to A
  4) A to B, bottom road, and back using the top road to A
  5) A to C, then back again to A.
- Also B to C has 8 routes:
  1) B to B clockwise, then down the left hand road to C
  2) B to B clockwise, then down the middle road to C
  3) B to B clockwise, then down the right hand line to C
  4) B to B anticlockwise, then left hand road to C
  5) B to B anticlockwise, then middle road to C
  6) B to B anticlockwise, then right hand road to C
  7) B to A, top road, and on to C
  8) B to A, bottom road and on to C.

Go through each of the other points and you should finish up with the two-stage route matrix shown in Figure M.2b).

The other way to find the two-stage route matrix is to *square* the one-stage route matrix, that is to multiply it by itself.

Three-stage route matrices

Similarly you can find the three-stage route matrix by *cubing* the one-stage route matrix.

The type of question you are likely to get will usually give you a route and ask you to calculate various matrices from this route. The most common errors will tend to be careless ones, where the candidate has not seen all the routes. Maybe a loop has not been round twice (clockwise and anticlockwise) or maybe even one point has been overlooked while looking for the two-stage routes. The best way to generally find any two-stage routes, is to square the one-stage route matrix, or at least to calculate the part necessary for the question.

TRANSFORMATION MATRICES

The majority of transformations that are considered in a GCSE examination will be of the type that can be represented by a (2 by 2) matrix:

- **Reflection** in the x axis will be
  $\begin{pmatrix} 1 & 0 \\ 0 & -1 \end{pmatrix}$
- **Reflection** in the y axis will be
  $\begin{pmatrix} -1 & 0 \\ 0 & 1 \end{pmatrix}$

- **Reflection** in the line $y = x$ will be
$\begin{pmatrix} 0 & 1 \\ 1 & 0 \end{pmatrix}$
- **Rotation** of 90° clockwise around the origin will be
$\begin{pmatrix} 0 & 1 \\ -1 & 0 \end{pmatrix}$
- Rotation of 90° anticlockwise around the origin will be
$\begin{pmatrix} 0 & -1 \\ 1 & 0 \end{pmatrix}$
- Rotation of 180° around the origin will be
$\begin{pmatrix} -1 & 0 \\ 0 & -1 \end{pmatrix}$
- **Enlargement** of scale factor K with the centre of enlargement the origin will be
$\begin{pmatrix} K & 0 \\ 0 & K \end{pmatrix}$
- **Shear** of scale factor K with the y axis invarient is
$\begin{pmatrix} 1 & 0 \\ K & 1 \end{pmatrix}$
- **Shear** of scale factor K with the x axis invarient is
$\begin{pmatrix} 1 & K \\ 0 & 1 \end{pmatrix}$
- One-way **stretch** from the x axis with stretch factor K, will be
$\begin{pmatrix} K & 0 \\ 0 & 1 \end{pmatrix}$
- One-way **stretch** from the y axis with stretch factor K, will be
$\begin{pmatrix} 1 & 0 \\ 0 & K \end{pmatrix}$

Position vectors of the transformed shape

These 2 by 2 transformation matrices multiply to matrices made up from the position vectors of the vertices of a shape. For example, a transformation is defined by:

$$T: \begin{pmatrix} x \\ y \end{pmatrix} \rightarrow \begin{pmatrix} 2 & 0 \\ 0 & 3 \end{pmatrix}\begin{pmatrix} x \\ y \end{pmatrix}$$

Use this to transform the triangle in Figure M.3 and describe fully the tranformation T.

The matrix that contains all the position vectors of the vertices of the triangle is:

$$\begin{pmatrix} 3 & 1 & 0 \\ 0 & 3 & 1 \end{pmatrix}$$

Hence to find the *position vectors* of the tranformed shape, we multiply the matrices as:

$$\begin{pmatrix} 2 & 0 \\ 0 & 3 \end{pmatrix} \begin{pmatrix} 3 & 1 & 0 \\ 0 & 3 & 1 \end{pmatrix} = \begin{pmatrix} 6 & 2 & 0 \\ 0 & 9 & 3 \end{pmatrix}$$

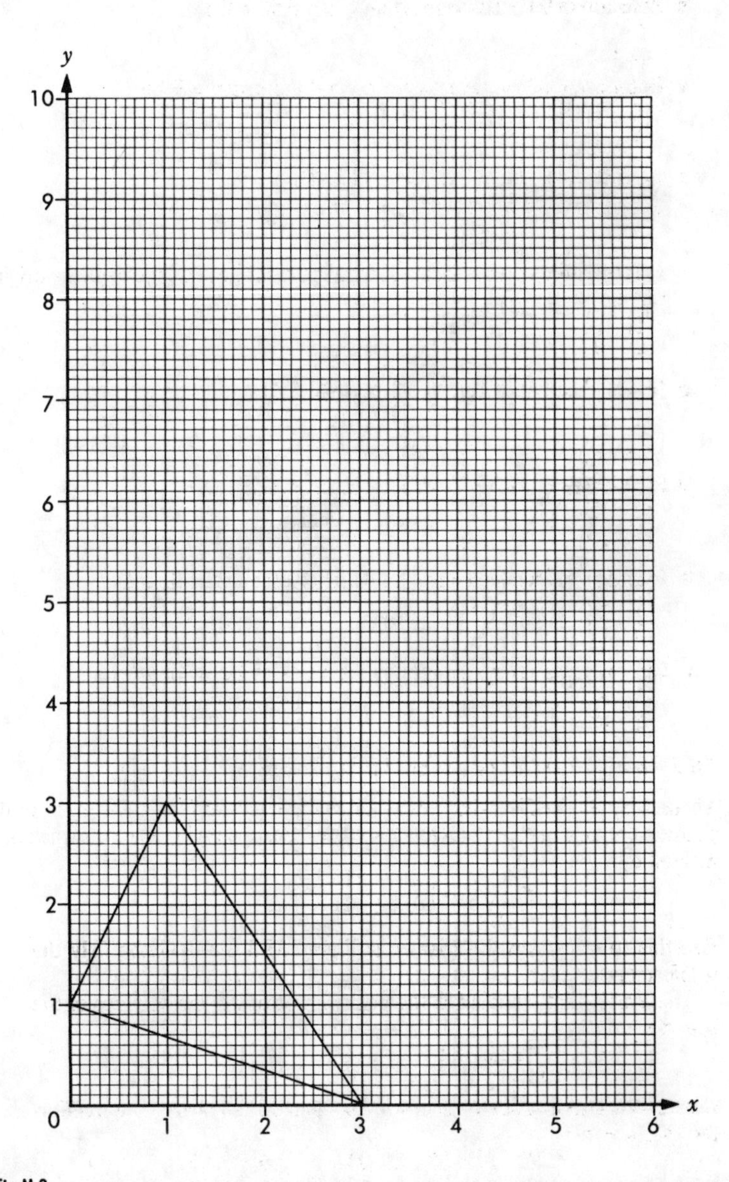

Fig M.3

We now draw this transformed shape out, as well as the original (Fig M.4). We see that the transformation is a 2-way stretch factor 2 from the y axis and stretch factor 3 from the x axis.

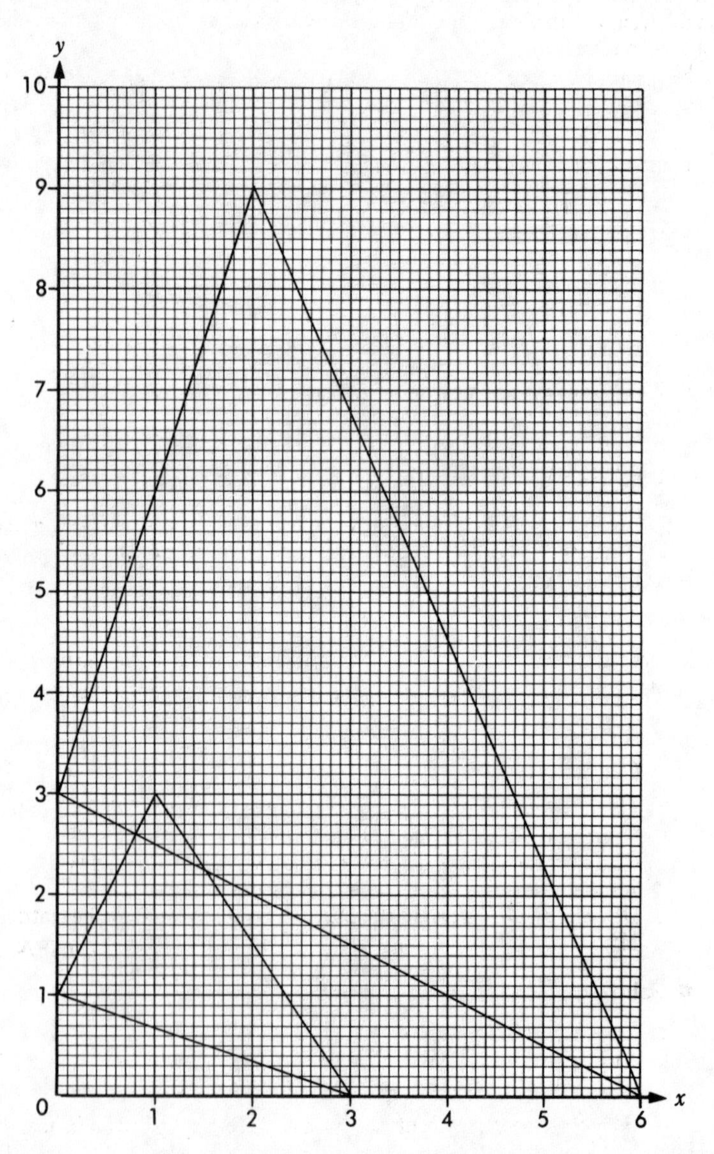

Fig M.4

Combined transformations

If we have two or more transformations **combined**, for example a reflection in the line $y = x$ followed by a rotation of 90° clockwise, then the transformation matrix of the combined transformation can be found by multiplying the two individual transformation matrices together (in the order of their operation).

For example, if we consider the combined transformation as above, where:

reflection in $y = x$ is given by $\begin{pmatrix} 0 & 1 \\ 1 & 0 \end{pmatrix}$

rotation of 90° clockwise is given by $\begin{pmatrix} 0 & 1 \\ -1 & 0 \end{pmatrix}$

then the *combined transformation* is found by:

$$\begin{pmatrix} 0 & 1 \\ -1 & 0 \end{pmatrix}\begin{pmatrix} 0 & 1 \\ 1 & 0 \end{pmatrix} = \begin{pmatrix} 1 & 0 \\ 0 & -1 \end{pmatrix}$$

Notice the order in which the transformation matrices are written down; the *right hand* matrix is the first transformation to be applied, the *left hand* matrix is the final transformation be be applied.

Hence the combined transformation matrix is given by: $\begin{pmatrix} 1 & 0 \\ 0 & -1 \end{pmatrix}$

This is a reflection in the x axis.

The most common errors in these types of question is to get the *signs* wrong. You will then of course get the tranformations the wrong way round and get the wrong combined transformation.

- **Exam Questions**

The matrix $M = \begin{pmatrix} -1 & 1 \\ -1 & -1 \end{pmatrix}$ represents a transformation T of the plane.

The image (x',y') of the point (x,y) under the tranformation T is given by

$$\begin{pmatrix} x' \\ y' \end{pmatrix} = \begin{pmatrix} -1 & 1 \\ -1 & -1 \end{pmatrix}\begin{pmatrix} x \\ y \end{pmatrix}$$

a) On the axes in Figure M.5, draw and label the image $OA'B'C'$ of the square $OABC$ under the transformation T.

b) The transformation T consists of an enlargement followed by a rotation. Describe these two transformations fully.

c) Describe the transformation of the plane represented by the matrix M^2.

(NEA; H)

- **Solution**

a) You can see from the base vectors that:

$\begin{pmatrix} 1 \\ 0 \end{pmatrix} \to \begin{pmatrix} -1 \\ -1 \end{pmatrix}$ and $\begin{pmatrix} 0 \\ 1 \end{pmatrix} \to \begin{pmatrix} 1 \\ -1 \end{pmatrix}$, also $\begin{pmatrix} 0 \\ 0 \end{pmatrix} \to \begin{pmatrix} 0 \\ 0 \end{pmatrix}$.

So we just need to substitute $\begin{pmatrix} 1 \\ 1 \end{pmatrix}$ into the equation and find it maps to $\begin{pmatrix} 0 \\ -2 \end{pmatrix}$, then we plot the shape as in Figure M.6.

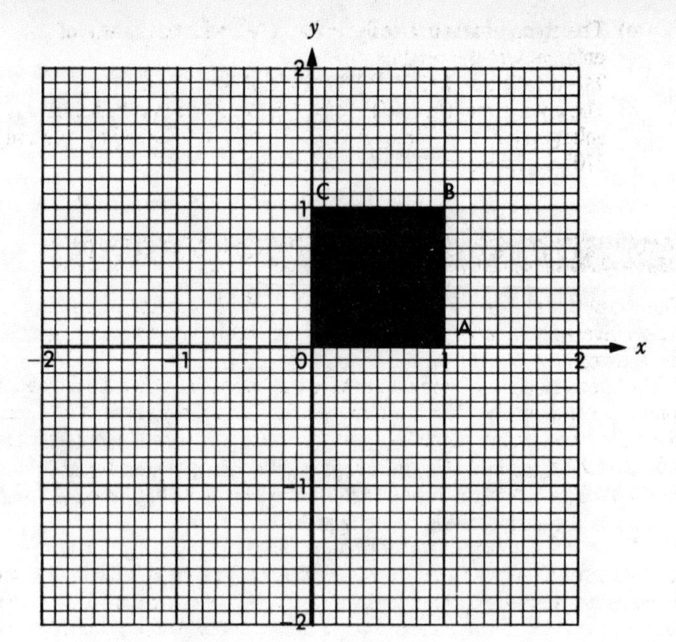

Fig M.5

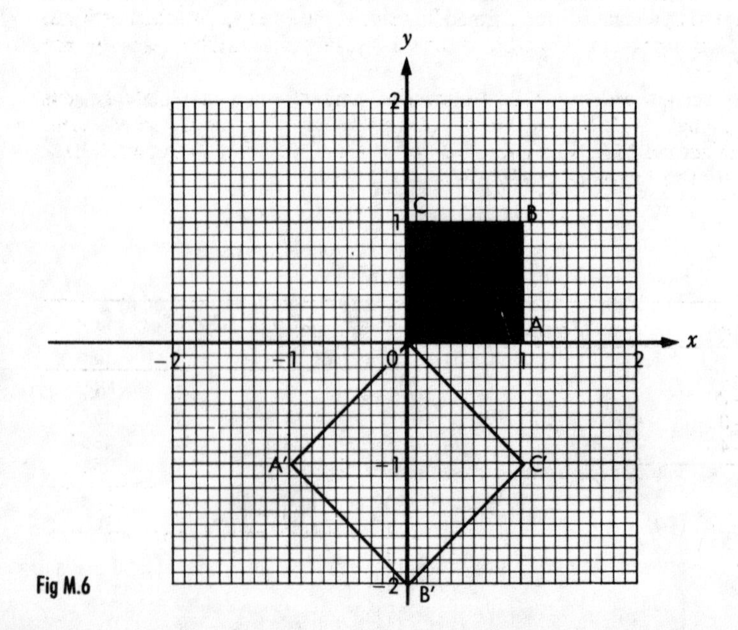

Fig M.6

b) The enlargement is a scale factor of $\sqrt{2}$ with the centre of enlargement the origin.
The rotation is of 135° clockwise about the origin.

c) The enlargement would then be $\sqrt{2}\times\sqrt{2}$ which is 2, hence enlargement scale factor 2 about the origin, followed by rotation of 270° clockwise (or 90° anti-clockwise).

MEAN

The mean usually refers to the *arithmetic mean,* which is the most common type of average. In fact it is usually what people are referring to when they say the average.

The mean of a set of numeric data is the sum of all the data divided by the total frequency of the data. In other words, it is the sum of all the numbers divided by how many numbers there are. For example, to find the mean of 4,6,3,2,7,12,8,5,13 and 9, you add up all the numbers to give you 69. Then divide by the number of numbers, which is 10, to give a mean of 6.9.

Estimated mean

An estimated mean is found from a *grouped* frequency distribution where we do not know all the individual items of data. We only know how many items of data are between certain limits. For example, consider the grouped frequency in Table (i) Figure M.7). This shows the number of candidates obtaining various ranges of marks in an examination. To *estimate* the mean, we assume that each candidate scored the middle mark in each range. We then estimate the total marks for that group by multiplying the middle mark by the frequency.

We can now add up all these estimates to find how many the total data adds up to. Then divide this by the total frequency to obtain our estimated mean.

The second table (ii) shows how we have done this in this example. Here we estimate the mean as $892 \div 37 = 24$.

Marks	*Frequency*
0 – 10	3
11 – 20	7
21 – 30	19
31 – 40	8

Marks	*Midway* *m*	*Frequency* *f*	*m × f*
0 – 10	5	3	15
11 – 20	15.5	7	108.5
21 – 30	25.5	19	484.5
31 – 40	35.5	8	284
	totals	37	892

Fig M.7 i) ii)

■ Exam Questions

A survey was carried out at a clinic during part of one day to see how long it took nurses to weigh and assess babies. The recorded times to the nearest minute, were as follows:

6	24	11	12	7
18	13	16	17	13
17	19	23	8	19
18	9	17	17	13
16	16	12	12	21

a) i) Calculate the mean time taken by the nurses to weigh and assess a baby on this day at the clinic.

ii) On average, how many babies could be assessed and weighed by one nurse during a three-hour session?

b) The clinic usually has around 80 babies to weigh and assess each day in a morning session between 9am and 12 noon. Using your answers to part a), calculate the number of nurses needed for each morning session.

(NEA; I)

■ Solution

a) i) Add up all the times to calculate the mean as $374 \div 25 = 14.96 = 15$ minutes.

ii) $3 \times 60 = 180$ minutes, then $180 \div 15 = 12$ babies per session.

b) $80 \div 12 = 6.67$ hence 7 nurses would be needed.

MEDIAN

The median is another type of average. It is the middle item of data, once that data has been sorted into an order of size. If there are two items of data in the middle, as there will be with an even number of items, then we add them together and divide by 2 to calculate the median. In general, if there are N numbers in a frequency distribution, then the middle item is the $\frac{(N+1)}{2}$ th.

For example, here are 15 test results. What is the median score?

(81, 63, 59, 71, 36, 99, 56, 31, 5, 65, 46, 83, 71, 53, 15)

Put the marks into order (5, 15, 31, 36, 46, 53, 56, 59, 63, 65, 71, 71, 81, 83, 99). Now find the middle one, which is 59. So the median score is 59.

Estimated median

The estimated median for a frequency distribution can be found by using a cumulative frequency graph. You can then read off what the middle item of data would be.

◀ Cumulative frequency ▶

MENSURATION

Mensuration is the study of the rules of measuring. Within any GCSE course you will be examined on how well you can measure lengths, areas and volumes.

This will include knowing simple area and volume formulae; you will also need to know *where to find* other formula on the *formula sheet* that you will be presented with in the examination. Of course the more formula you can learn off by heart, instead of having to use the formula sheet, the easier and quicker you will be able to answer the examination questions.

You also need to be able to *recognise* those situations where you can use a right-angled triangle to find lengths by either **trigonometry** or by the theorem of **Pythagoras**. If an *angle* is given, use trigonometry; whereas if *two sides* are given, then use *Pythagoras*.

■ Exam Questions

1 A circle is contained within a square as shown in Figure M.8. The radius of the circle is 5 cm. Estimate the area of the shaded part of the diagram. (Take the value of π to be 3)

(SEG; I)

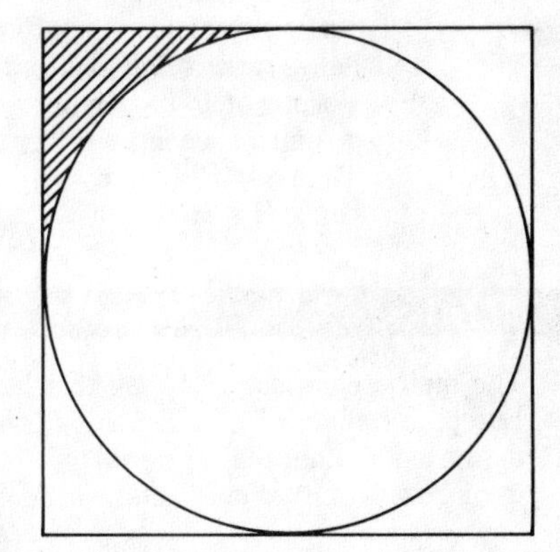

Fig M.8

2 Figure M.9 shows two closed cylindrical cans, *A* and *B*. The radius of *A* is 4 cm and its height is 12 cm. The radius of *B* is 8 cm and its height is 6 cm.

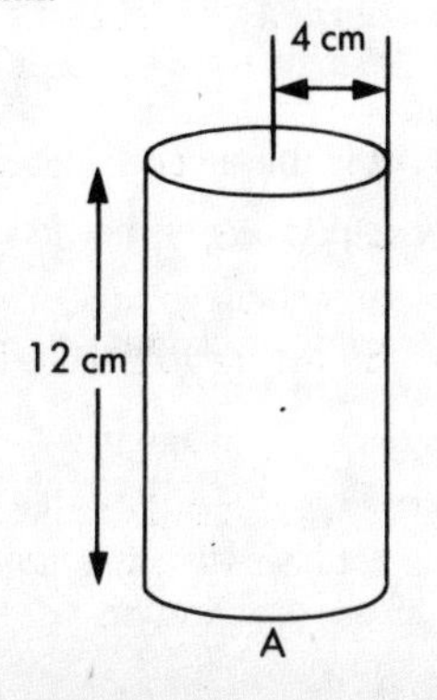

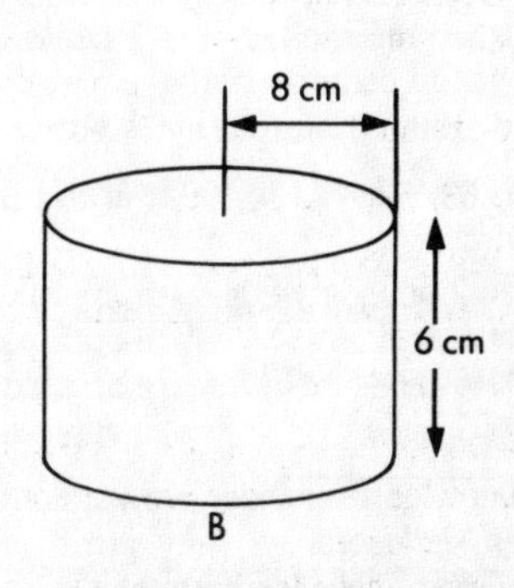

Fig M.9

a) Find, in the form 1:n, the ratio
   i) volume of A: volume of B,
   ii) total surface area of A: total surface area of B.

b) Two cylinders have the same volume. The first has radius r and height h. If the radius of the second is $2r$, find its height in terms of h.

(MEG; H)

■ Solutions

1 The area of the rectangle = $10 \times 10 = 100\text{cm}^2$
the area of the circle = $3 \times 25 = 75\text{cm}^2$
shaded area = $(100-75) \div 4 = 6.25\text{cm}^2$

2 a) i) Volume of A = $12\pi 16 = 192\pi$
volume of B = $6\pi 64 = 384\pi$
hence ratio of A:B = 192:384 = 1:2

ii) Surface area of A = $\pi \times 8 \times 12 + 2 \times \pi \times 16 = 128\pi$
surface area of B = $\pi \times 16 \times 6 + 2 \times \pi \times 64 = 224\pi$
hence ratio of A:B = 128:224 = 1:1.75

b) Volume of first = $\pi r^2 h$
volume of second = $\pi(2r)^2 H$.
Then $\pi 4r^2 H = \pi r^2 h$
hence $H = \dfrac{\pi r^2 h}{4\pi r^2} = \dfrac{h}{4}$

METRIC

The metric units are the ones that have been used throughout the rest of Europe for quite a while. Britain is trying to encourage its population to use metric units rather than **imperial** units. You ought to know and be familiar with the following metric unit facts:

- 1 kilogram = 1000 grams or 1kg = 1000g
- 1 kilometre = 1000 metres or 1km = 1000m
- 1 kilowatt = 1000 watts or 1kW = 1000w

(It is worth remembering that *kilo* means 1000.)

Other metric facts you ought to be aware of:

- 1000 kilograms = 1 tonne or 1000 kg = 1 t
- 10 millimetres = 1 centimetre or 10 mm = 1 cm
- 100 centimetres = 1 metre or 100 cm = 1 m
- 1000 millilitres = 1 litre or 1000 ml = 1 l

It is very useful to be aware of the *rough equivalents* of metric and imperial units:

- 2 pounds weight are approximately equal to 1 kilogram
- 3 feet are approximately equal to 1 metre
- 5 miles are approximately equal to 8 kilometres
- 1 gallon is approximately equal to 4.5 litres

A knowledge of these conversion factors is not needed, but a *familiarity* with them is useful as they could crop up in an examination question. Such familiarity will also be useful in the real world of work. ◀ Imperial ▶

MIXED NUMBERS

A mixed number is one which is written with two distinct parts; one part includes a *whole number*, while the other part is the *vulgar fraction*. For example 5⅔ is a mixed number.

To change a mixed number to a **decimal** number, start with the vulgar fraction. Then, using your calculator, divide the numerator (top number) by the denominator (bottom number); add to your result the whole number part of the mixed number.

For example, to change 4⅜ to a decimal we would use the calculator to divide 3 by 8 to get 0.375, then add 4 to this, giving us 4.375. This is the decimal equivalent of 4⅜.

MODE

The mode is a type of average; it is what most people have. The mode is the value occurring most times in a particular frequency distribution. The number 1 hit single each week is the record that has sold more copies than any other record that week; it is therefore the mode record or, as it is sometimes called, the *modal* record.

- In a **bar chart**, the mode is always the data represented by the longest bar.
- In a **histogram**, the mode will be the data represented by the bar of greatest area.
- In a **pie chart**, the mode will be the data represented by the largest sector (the largest angle).

MULTIPLES

The multiples of an **integer**, say N, are the integers that will divide exactly by N. In other words, they will be the results from the N-times table. For example, the multiples of 5 are 5, 10, 15, 20, 25, etc.

Common multiples

The common multiples of two or more integers are those integers that can be divided exactly by these integers. For example, the common multiples of 5 and 3 are 15, 30, 45 etc., as these numbers can be divided exactly by *both* 3 and 5.

Lowest common multiple

The **lowest common multiple** (LCM) is the smallest of the common multiples. In the example above the LCM is 15.

MULTIPLYING/MULTIPLICATION

◀ Directed numbers, Matrices ▶

NEGATIVE ENLARGEMENTS

◀ Enlargements ▶

NEGATIVE NUMBERS

Negative numbers are those found on a thermometer below freezing point. They have a *negative* (minus) *sign* to go with them, so that we know they are negative numbers.

Any number smaller than 0 is a negative number.

◀ Directed numbers ▶

NEGATIVE SCALE FACTOR

◀ Enlargements ▶

NETS

A net is a flat shape that can be folded up to create a solid shape (Fig N.1). You need to be able to *recognise* what shape a net will fold into. You must also be able to *draw* a net for any of the usual, regular, three-dimensional shapes; for example **cubes, cuboids, tetrahedrons, pyramids, cylinders** and **cones.**

The question may give you a *net* and ask you to identify or draw the *shape* it will make, or it may ask you to draw a sketch of a *net* for some particular *shape*.

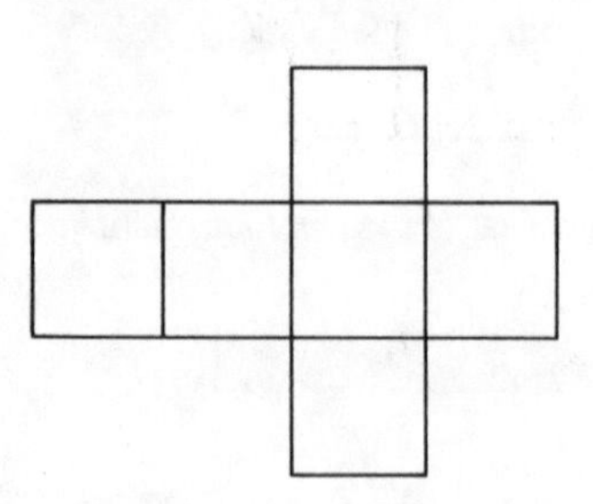

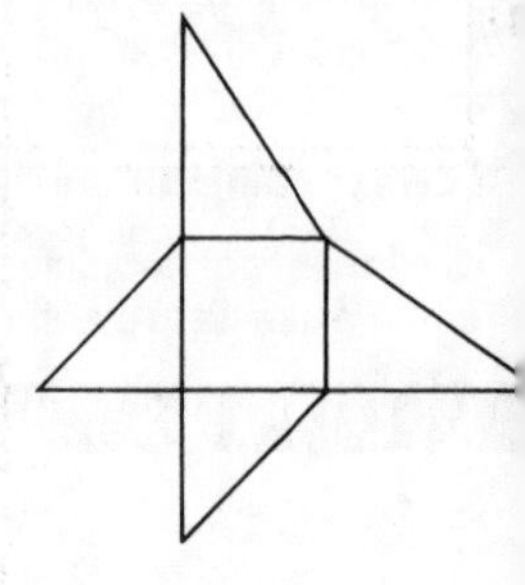

Fig N.1 Nets

The most common mistake is for candidates to fail to visualise the three-dimensional shape that a net makes, and to answer the question with a 2-dimensional shape. There are of course, for each solid shape, quite a few different ways to draw the net; all correct but different.

- **Exam Question**
  The closed packet sketched in Figure N.2 measures 30cm by 18cm by 7cm. Sketch a net for this packet, marking in the dimensions.
  (NISEC;B)

- **Solution**
  See Figure N.3; this is not a unique answer, and there are a number of different possibilities.

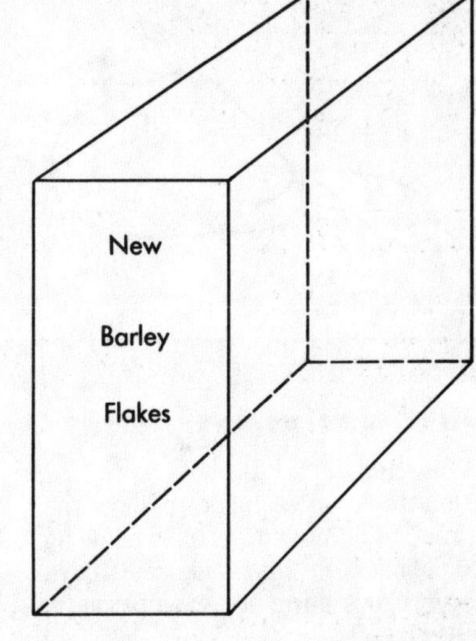

Fig N.2

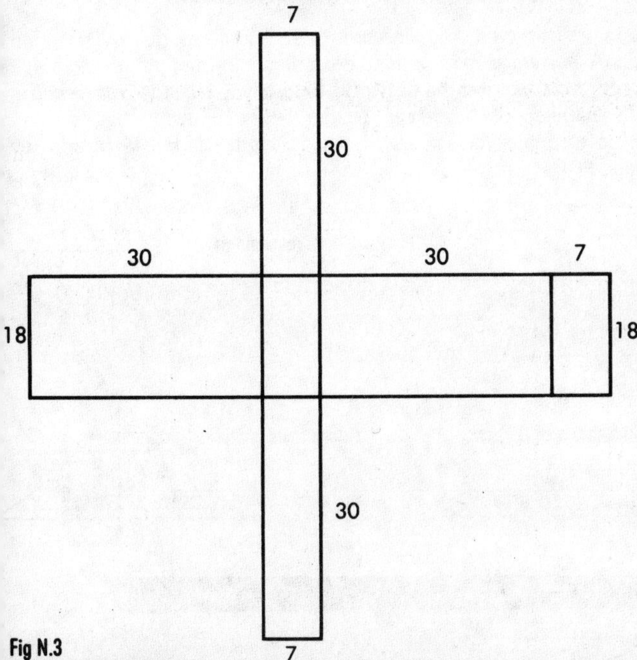

Fig N.3

NETWORKS

A network is a collection of points (**nodes**), connected by lines (roads). Some examples are shown in Figure N.4.

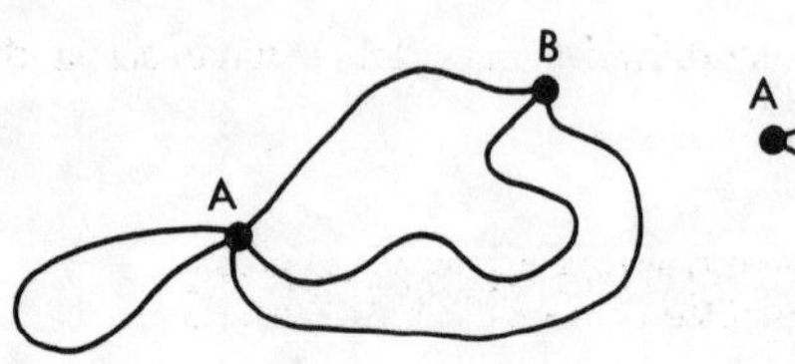

Fig N.4 Networks

Directed network

A directed network is a network that has arrows on each line to show the direction of movement allowed along each line. Note that a line could have two arrows on it, for example as in Figure N.5.

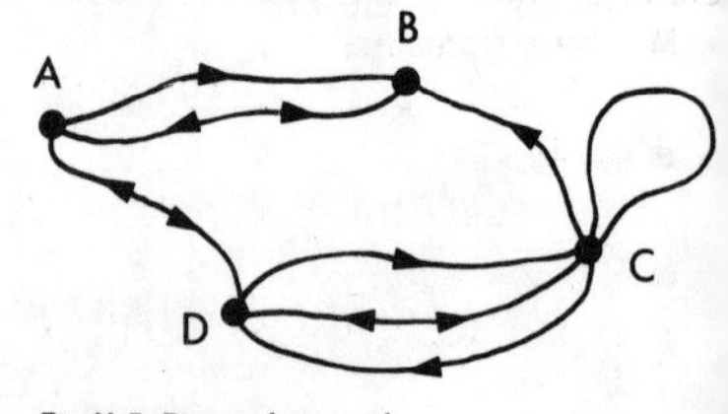

Fig N.5 Directed network

For any network, directed or undirected, there is a **matrix** that will describe that network. A *one-stage route matrix* describes the number of one-stage routes between each point on the network. Similarly a *two-stage route matrix* describes the number of two-stage routes between each point on the network. Some examples below are shown in Figure N.6. ◀ **Matrix** for route matrices ▶

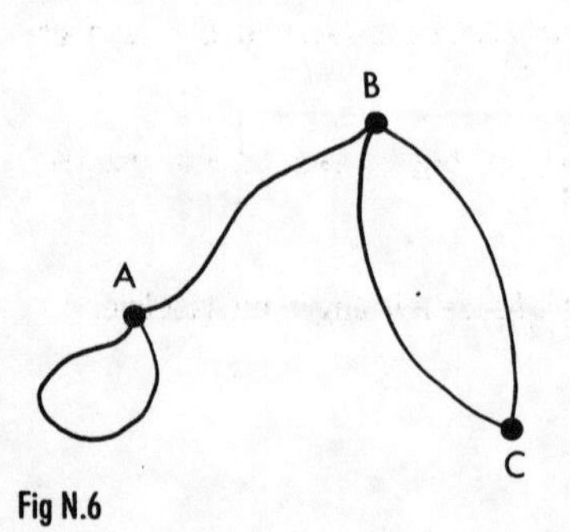

Fig N.6

1-stage route matrix

		To A	To B	To C
From	A	2	1	0
	B	1	0	2
	C	0	2	0

2-stage route matrix

		To A	To B	To C
From	A	5	2	2
	B	2	5	0
	C	2	0	4

NODES

◀ Networks ▶

NONAGON

A nonagon is a nine-sided polygon.

NULL

This is a term often used with **hypothesis**. The Null Hypothesis is the hypothesis (or theory) to be tested.

◀ **Matrix** for null matrix ▶

NUMBER PATTERNS

Both in coursework and in your final end of course examination, you will be expected to investigate, work out and recognise a variety of number patterns. Some of these will be based on the following:

- **Prime numbers**
  2, 3, 5, 7, 11, 13, 17, 19, 23, 29...
- **Multiples**
  3, 6, 9, 12, 15, 18, 21, 24, 27, 30...
- **Square numbers**
  1, 4, 9, 16, 25, 36, 49, 64, 81, 100...

Searching for patterns

The most common way is to look at the *differences*. This will in fact help you to find most of the patterns and then to continue them. For example, find the next three numbers in the following number pattern:

3, 7, 11, 15, 19...

Looking at the *differences* we see that the difference is four each time; so the pattern can be continued by simply adding on 4 each time, to give 23, 27 and 31.

Here is another example. Find the next three numbers in the number pattern:

8, 9, 11, 14, 18, 23...

Looking at the *differences* we see that they get bigger each time by one, so the next three numbers will be 29, 36 and 44.

- **Exam Question**
  Look at the following pattern. There is a figure missing in the last line.

$$1 \times 1 = 1$$
$$11 \times 11 = 121$$
$$111 \times 111 = 12321$$
$$1111 \times 1111 = 1234321$$
$$11111 \times 11111 = 1234{*}4321$$

a) What figure should * be?
b) Complete the answer for this line of the pattern.
$111111 \times 111111 =$
c) Write down the next complete line of the pattern.

■ **Solution**

a) 5
b) 12345654321
c) 1111111 × 1111111 = 1234567654321

NUMBERS

Numbers may take a variety of forms.

Irrational numbers

An irrational number is one that *cannot* be expressed as a vulgar fraction, i.e. as a/b, where both a and b are integers (whole numbers). It is also true to say that any irrational number *cannot* be expressed as either a **terminating decimal** or a **recurring decimal**.

The most commonly quoted examples of irrational numbers are π and $\sqrt{2}$

Prime numbers

A **prime number** is an integer that has exactly two **factors**; these two factors will of course always be the number itself and 1.

The first few prime numbers are 2, 3, 5, 7, 11, 13, 17, 19, 23, 29 and 31. You ought to be familiar with the numbers which are prime numbers and to note that 1 is *not* a prime number as it does not have two factors, only one.

Rational numbers

A **rational number** is a number that *can* be expressed as a vulgar fraction, i.e. as a/b, where a and b are both integers. It is true to say that every rational number is either a **terminating decimal** or a **recurring decimal**. The vast majority of numbers are rational.

Real numbers

A **real number** is any number from the *union* of two sets, namely rational and irrational numbers. Every number that you come across in a GCSE course or examination will be a real number.

Square numbers

A **square number** is a number that can be formed by multiplying a whole number by itself. For example, 16 is a square number because 4 multiplied by itself is 16.

The first ten square numbers 1, 4, 9, 16, 25, 36, 49, 64, 81, 100 ought to be known.

There will be few questions that test specific knowledge of the above numbers; they are more likely to be examined *within* a question.

◀ Directed number ▶

OCTAGON

An octagon is an eight-sided **polygon**. A *regular* octagon will have eight **lines of symmetry**, and **rotational symmetry** of order eight.

OGIVE

The ogive is the special shape that you get on a cumulative frequency curve. On the cumulative curve in Figure O.1 the curved line is the *ogive*. From it you could evaluate the **quartiles** and the **median** of this distribution.

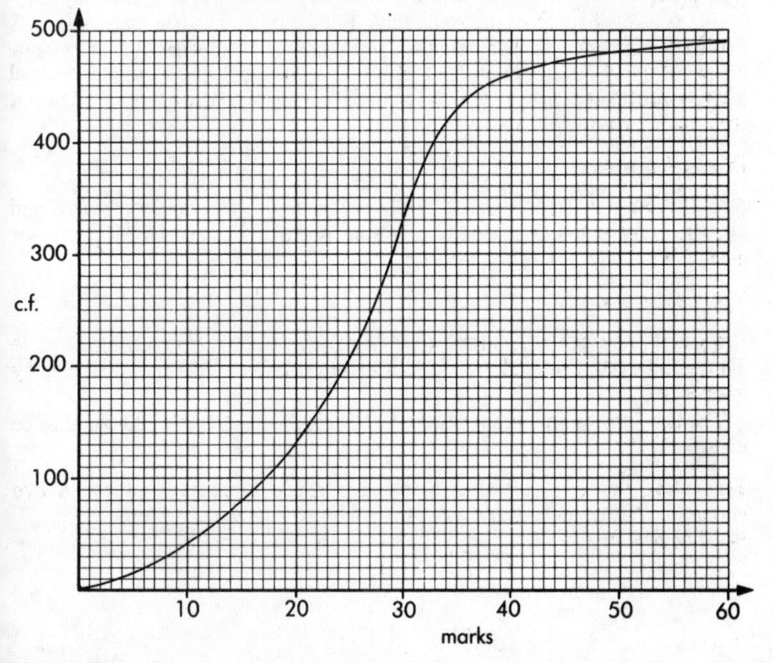

Fig O.1 Ogive

OPPOSITE

Opposite is the word given to that side of a right-angled triangle opposite the *angle* that is being *calculated* by **trigonometry** (or has been *given*, and where you are using trigonometry to calculate a side) (Fig O.2).

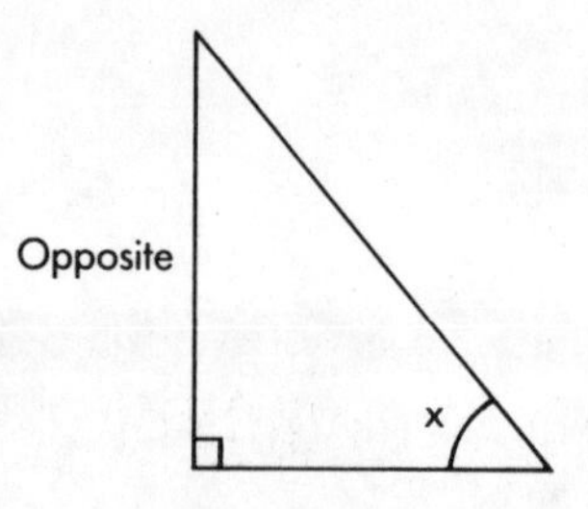

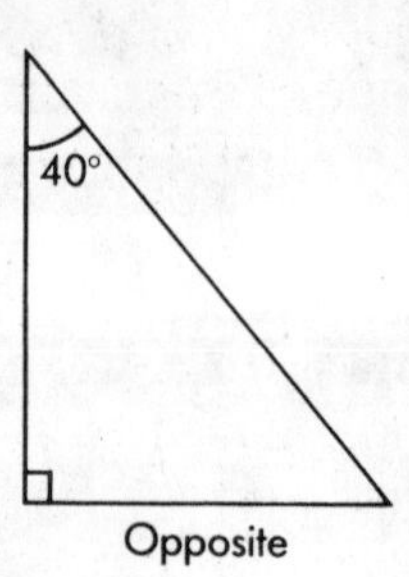

Fig O.2

OR RULE

The 'OR' rule comes from the topic of **probability**. To find the probability of event A *or* event B, the probability of each event is *added* together.

However these two events must be *mutually exclusive*, that is to say they cannot possibly happen at the same time. For example, in a bag that contains 5 toffees, 4 jellies and 3 mints, what is the probability of selecting one at random and it being a jelly *or* a mint?

The probabilities are:

Selecting a jelly.............. $\frac{4}{12}$
Selecting a mint .... $\frac{3}{12}$
add these two together to get $\frac{7}{12}$

So probability of selecting a jelly *or* a mint $= \frac{7}{12}$.

PARALLEL

Two lines are parallel if the **perpendicular** difference between them is always the same. Parallel lines are usually thought of as straight lines, but do not have to be. Two *circles* of different radii but the same centre will be parallel with each other, but they are not straight lines. Figure P.1 gives some examples of parallel lines.

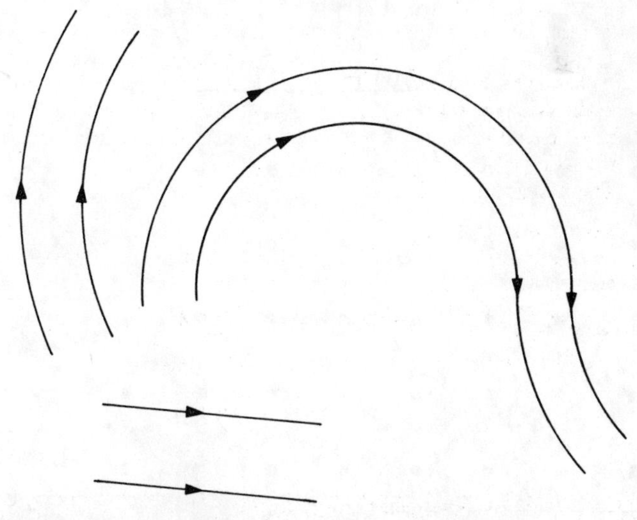

Fig P.1 Parallel lines

PARALLELOGRAM

A parallelogram has four sides and the opposite sides are of equal length, as in Figure P.2. The opposite sides are parallel. In a parallelogram any two angles *next to each other* will always add up to 180°.
For example, $a + b = b + c = c + d = d + a = 180°$.
Also, the angles *opposite each other* will be equal. For example, a = c, b = d.

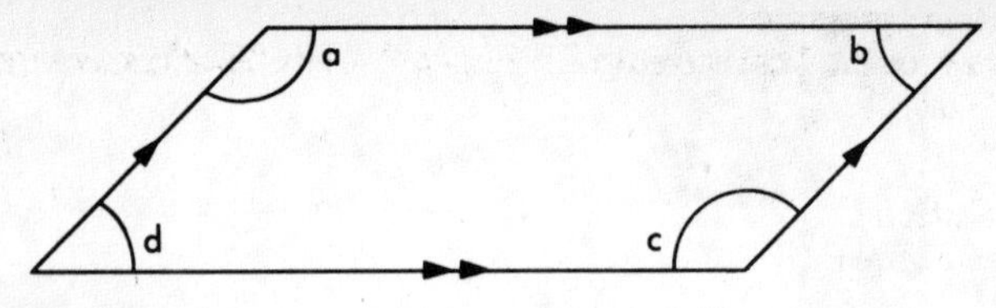

Fig P.2 Parallelogram

The *area* of a parallelogram is found by multiplying the base length by the perpendicular height.

The most common error in finding the area of a parallelogram is multiplying the base length by the slant height.

In order to find the area of the parallelogram in Figure P.3 you need to use **trigonometry** to calculate the perpendicular height of the parallelogram. Then multiply this perpendicular height by the base length.

Fig P.3

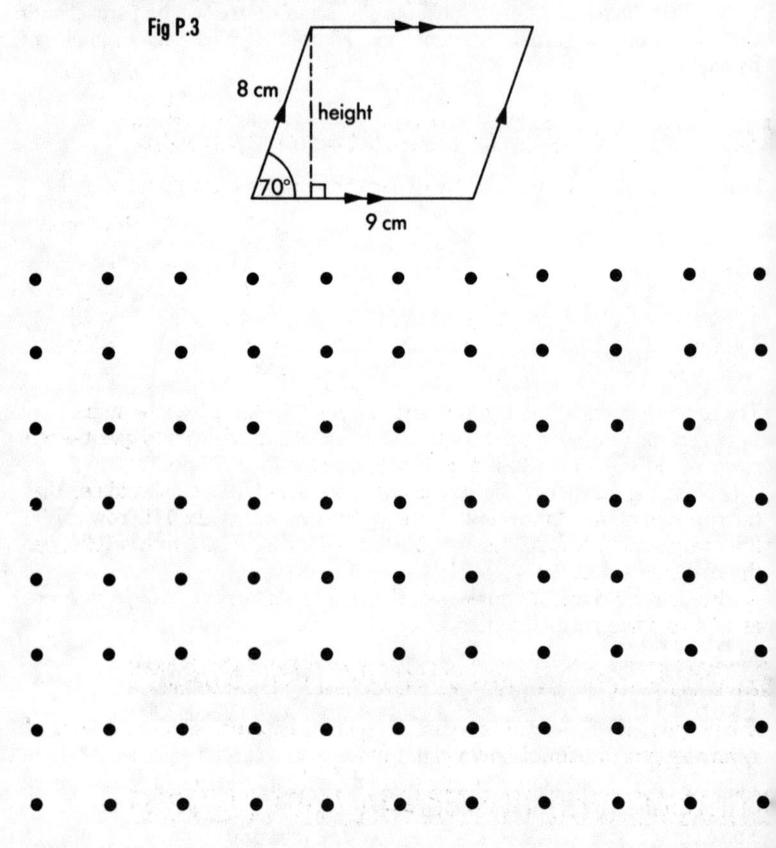

Fig P.4

■ **Exam Question**
Draw, on the 1 cm dotted grid in Figure P.4, a parallelogram of area 12 cm^2.

(NEA;I)

■ **Solution**
See Figure P.5

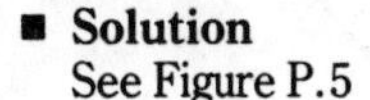

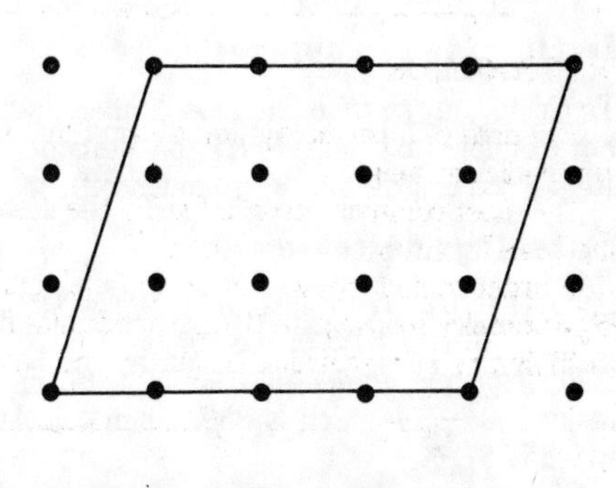

Fig P.5

PASCAL'S TRIANGLE

Pascal's triangle is a particular triangular array of numbers, as follows:

		Row sum	
1st row	1	= 1	$= 2^0$
2nd row	1 1	= 2	$= 2^1$
3rd row	1 2 1	= 4	$= 2^2$
4th row	1 3 3 1	= 8	$= 2^3$
5th row	1 4 6 4 1	= 16	$= 2^4$

Try to write down a) the 6th row and row sum, b) the 11th row sum.

You should be able to see how the pattern builds itself down to give the 6th row as $1 + 5 + 10 + 10 + 5 + 1$, with a row sum of $32 = 2^5$.

Look at the number of the row and the row sum, and you should see that the row sum of the nth row is 2^{n-1}. Hence the row sum of the 11th row will be 2^{11-1} which is 2^{10}. Now, 2^{10} is $2^5 \times 2^5$ which will be 32×32 which is 1024, i.e. the row sum will be $1024 = 2^{10}$.

Also, looking diagonally down the columns you will see the triangle numbers as well as some polyhedra numbers.

PENTAGON

A pentagon is a five-sided polygon. A *regular* pentagon will have 5 **lines of symmetry** and **rotational symmetry** of order 5.

PERCENTAGES

One percent, written as 1% means one out of one hundred, or 1/100 or 0.01.

Similarly 3% means 3 out of 100, or $\frac{3}{100}$, or 0.03, while 27% means 27 out of 100, or $\frac{27}{100}$, or 0.27.

For example, to find 35% of £8 we calculate... $8 \times \frac{35}{100} = £2.80$

PERCENTAGE CALCULATIONS

Fractions to percentages

To change any *fraction* into a percentage, simply multiply that fraction by 100. For example, to change $\frac{15}{35}$ into a percentage, calculate on your calculator... $15 \times \frac{100}{35}$ to give 42.9% (rounded off).

Percentage decrease

If we wish to *decrease* an amount by say 16%, then we really need to calculate (100 − 16)% or 84% of the amount.

For example, to decrease £9 by 14%, first recognise that this decrease is really 100 − 14 which is 86%; then calculate on your calculator $9 \times \frac{86}{100}$ which is £7.74.

Of course there is another way to do this which simply requires you to calculate 14% of £9 and then to *subtract* this from £9, but this is a much longer method.

Percentage increase

If we wish to *increase* an amount by say 8%, then we really need to calculate 108% of the amount.

For example, to increase £15 by 8% simply calculate on the calculator $15 \times \frac{108}{100}$ which gives £16.20.

The longer method would simply require you to calculate the 8% of £15 and then to *add* this onto the £15.

Percentage profit

If an article is bought at a cost price and then sold to make a profit, the percentage profit is the fractional increase multiplied by 100.

For example, Brian bought a Datsun for £250 and then the next day managed to sell it for £275. What was his percentage profit?

The profit was £275 − £250 which is £25; so the profit fraction is $\frac{25}{250}$. Multiply this by 100 to give 10%, so the percentage profit is 10%.

PERIMETER

The perimeter is the length of an outside edge of a plane shape. For example, the perimeter of a **rectangle** is all the four lengths added to each other. The perimeter of a **circle** is the circumference of that circle.

PERPENDICULAR

A **line** that is perpendicular to another line is at right angles to it. A **plane** that is perpendicular to another plane is at right angles to that plane.

Construct a perpendicular

◀ Construction ▶

PERSPECTIVE

The presenting of solid objects on a plane surface in such a way that they look like the *actual* objects viewed from particular points.

PICTOGRAM

A pictogram is a display of information using pictures to represent the **frequency**. Figure P.6 is a pictogram; it displays information with pictures. Here it displays the number of cups of tea drunk per week by Yorkshire Bank Managers in certain parts of Yorkshire.

Notice that the key is one cup of tea per 10 cups drunk; so we use half a cup of tea to represent 5 cups drunk.

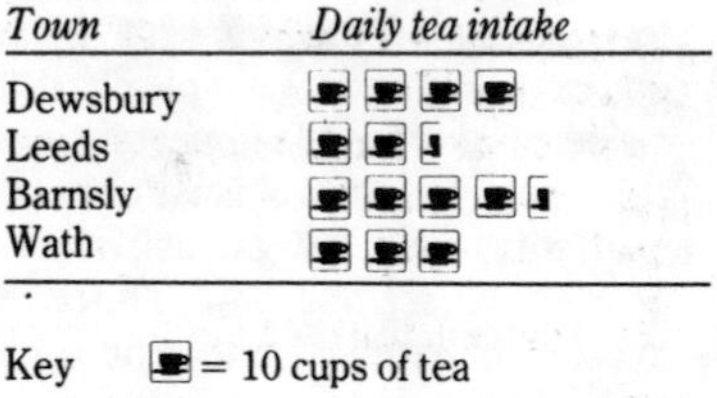

Town	*Daily tea intake*
Dewsbury	☕☕☕☕
Leeds	☕☕ ½
Barnsly	☕☕☕☕ ½
Wath	☕☕☕

Key ☕ = 10 cups of tea

Fig P.6 Pictogram

This type of display can be effective and more interesting than other types of display of data. However it tends to be rather less accurate; for example it would be difficult to represent one cup of tea on the pictogram shown.

PI (π)

Pi is the ratio found when you divide the **circumference** of a circle by its **diameter.** Its presence has been known for a long time. However its accuracy has troubled many mathematicians throughout history in that it is an **irrational number** and as such we cannot state it exactly.

Your calculator holds the value of pi to as accurate a level as you will need, and you are advised always to use the *calculator value of pi* whenever you need to use pi.

Examination questions will accept the use of **rounded off** values of pi, such as 3.14 or 3.142. When using any value of pi, do remember to round off to a suitable degree of accuracy.

Pi is particularly used in the following formulae:

- Circumference of circle = $\pi \times$ diameter of the circle
- Area of circle = $\pi \times$ square of the radius of the circle

PIECEWORK

Piecework is what we call the payment for each piece of work done. This means that some people are paid purely for the amount of work they actually do.

For example, James is paid 17p for every component he assembles on his production line. In one day he managed to assemble 138 components. How much would he earn in a week if he was able to do this each day in a 5 day week?

He would earn £0.17 × 138 = £23.46; so in 5 days he would earn £23.46 × 5, which totals £117.30

PIE CHART

A pie chart is a circular picture which is divided into the ratio of the frequencies of the different events occuring. Figure P.7 is a pie chart, so called since it has the appearance of a pie. This pie chart illustrates the transport used by pupils of High Storres School one day. The actual information is difficult to read accurately, but it does show us that the vast majority of pupils at the school come by bus.

You are quite likely to be asked in an examination question to extract information from a pie chart.

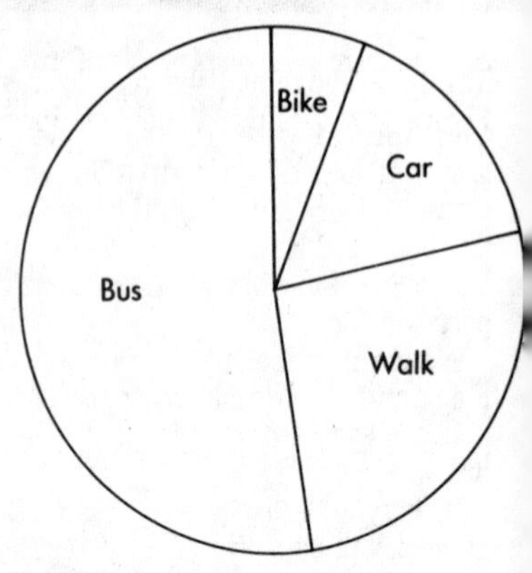

Fig P.7 Pie chart

CONSTRUCTING A PIE CHART

There is a set way to go about *constructing* a pie chart from given information. For example, suppose we are given the information in the table below:

Church Expenditure

Item	*cost £*
Electric gas	6,300
Minister wages and expenses	12,700
New Hymn books and Bibles	550
Posters and notices	280
Donations to other charities	3,800

We need to find the *angle of the sector* that each separate item will be. We do this by finding what fraction of the whole data each item is; we then use this to find the same fraction of a complete circle (i.e. of 360°).

The table below illustrates what we would do with the above information:

item	*cost*	*angle*
Electric/gas	6,300	$6300/23630 \times 360 = 96°$
Minister	12,700	$12700/23630 \times 360 = 193°$
Books	550	$550/23630 \times 360 = 8°$
Posters etc	280	$280/23630 \times 360 = 4°$
Donations	3,800	$3800/23630 \times 360 = 58°$
Totals	23,630	359°

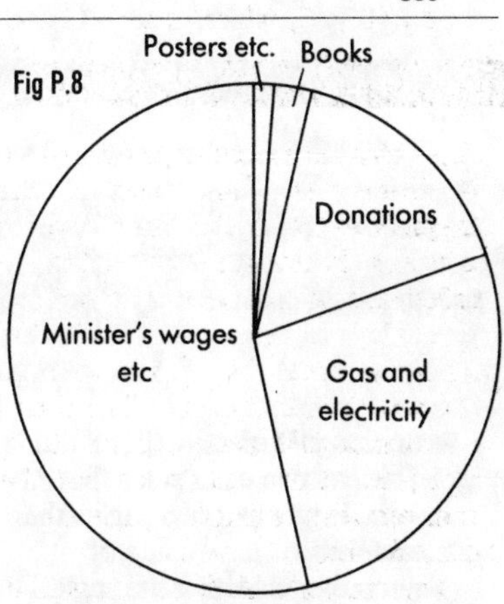

Fig P.8

The table above and the pie chart in Figure P.8 illustrate what we have done. Note how the angles in the chart have been rounded off to the nearest degree, so that their total is not exactly the 360° which you would normally expect to get. The pie chart was drawn smallest angle first, then next smallest, etc., until the largest one was drawn last of all. By drawing the largest angle last, any slight error will be less noticeable.

Note that although the pie charts you see in every day use will probably *not* have their sector angles labelled (e.g. 96°), this is usually expected in an examination question. In this case you are trying to show that you *know* what the angle should be.

■ **Exam Questions**

The pie chart in Figure P.9 is published by Dr Barnardo's in Northern Ireland to show how they spend each £1. If a school raised £2,680 for Dr Barnardo's;

a) How much of that money would go directly to child care?

b) How much would be spend on education and appeals?

(NISEC; I)

HOW WE USE EACH £1

3p necessary head office administration

1p films and information

16p education and appeals

80p directly to child care

ONE POUND

Fig P.9

■ **Solution**

a) $\dfrac{2680 \times 80}{100} = £2144$

b) $\dfrac{2680 \times 16}{100} = £428.80$

PLANE SHAPES

A plane shape is one that is two-dimensional; that is, it will have width and height, but no depth.

You should recognise, be able to name, and know the distinguishing features between the following plane shapes:

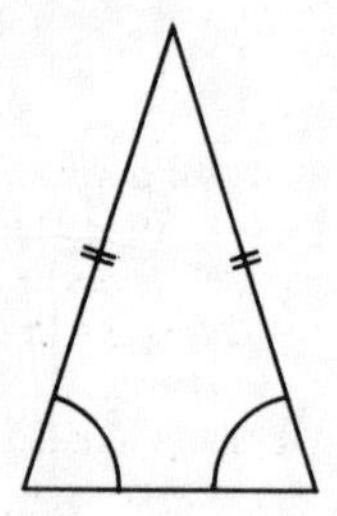

Fig P.10 Isosceles triangle

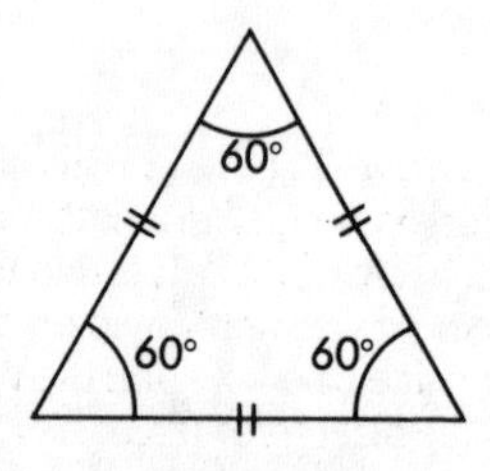

Fig P.11 Equilateral triangle

- An *isosceles* triangle (Fig P.10). This has two of its sides the same length and two angles the same. It will have a **line of symmetry** bisecting the angle included between the two equal sides.

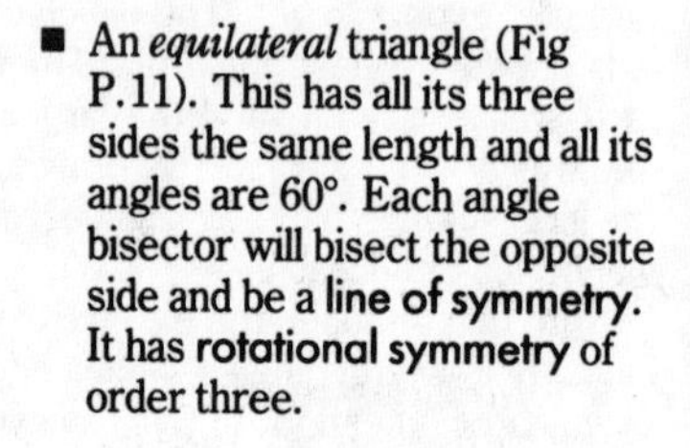

- An *equilateral* triangle (Fig P.11). This has all its three sides the same length and all its angles are 60°. Each angle bisector will bisect the opposite side and be a **line of symmetry**. It has **rotational symmetry** of order three.

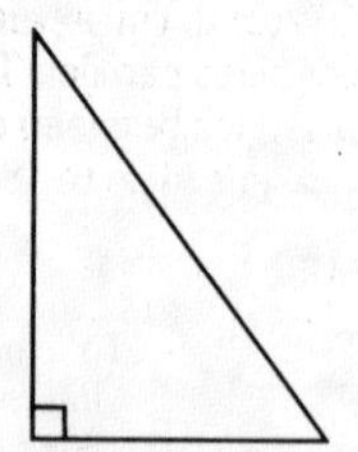

Fig P.12 Right-angled triangle

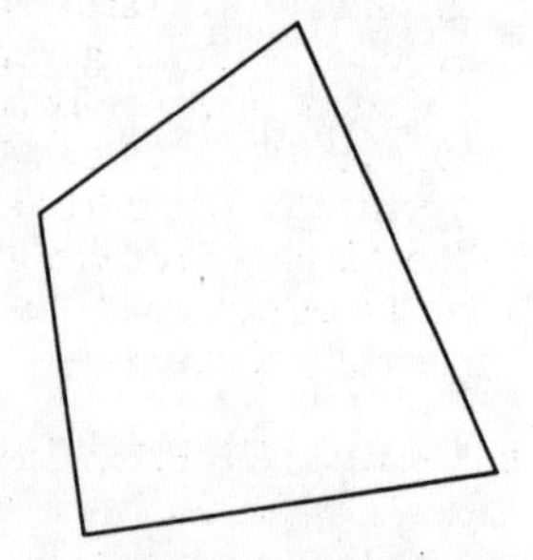

Fig P.13 Quadrilateral

- A *right-angled* triangle (Fig P.12). This is one that contains a right angle.
- A *quadrilateral* (Fig P.13). This has four sides, and the four angles it contains add up to 360°.

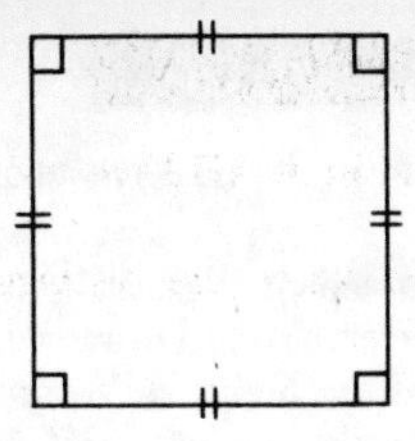

Fig P.14 Square

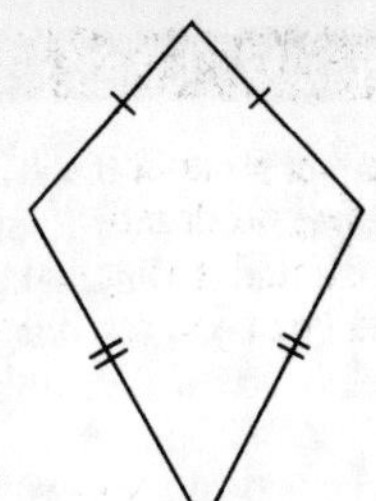

Fig P.15 Kite

- A *square* (Fig P.14). This has all four sides equal in length and each angle is 90°. It has four **lines of symmetry**, namely the two line bisectors and the two angle bisectors. It also has **rotational symmetry** of order four.

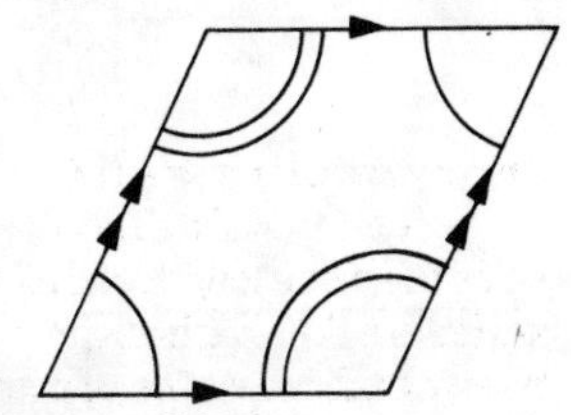

Fig P.16 Parallelogram

- A *parallelogram* (Fig P.16). This has four sides and the opposite two sides are equal in length as well as being parallel. The angles *next to each other* will add up to 180° and the angles *opposite each other* will be equal. There is no **line of symmetry** but it does have **rotational symmetry** of order two.

- A *rhombus* (Fig P.18). This is a parallelogram that has all its sides the same length. Its diagonals are perpendicular, and bisect each other. There are two **lines of symmetry**, namely the angle bisectors. It has **rotational symmetry** of order two.

- A *kite* (Fig P.15). This is recognisable as a kite shape. It has four sides as shown; the top two sides are the same length and the bottom two sides are the same length. There is one **line of symmetry**, namely the line that bisects both angles included between the equal sides.

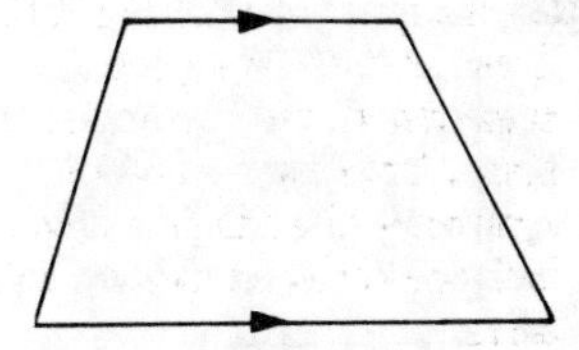

Fig P.17 Trapezium

- A *trapezium* (Fig P.17). This is a quadrilateral with a pair of opposite sides parallel. The pairs of angles between each parallel side add up to 180°.

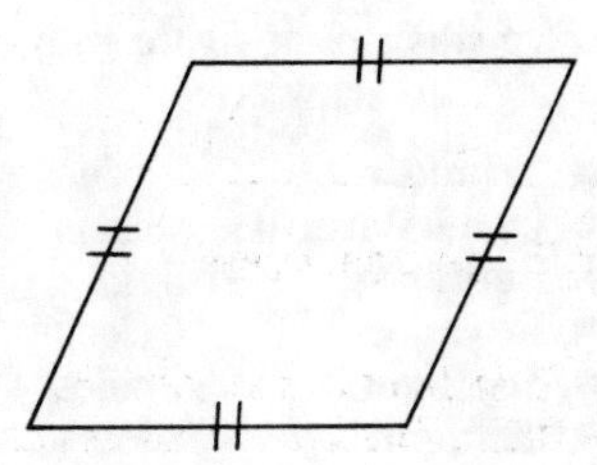

Fig P.18 Rhombus

PLANS

The plan of a three-dimensional shape is the view you get when looking down from directly above the shape.

For example, Figure P.19 shows some shapes and their plans.

The most common mistake when asked to draw a plan is to draw a diagram that still has that three-dimensional look of *perspective* about it. You must be clear that a plan is a *two-dimensional view only* and it must have a two-dimensional appearance.

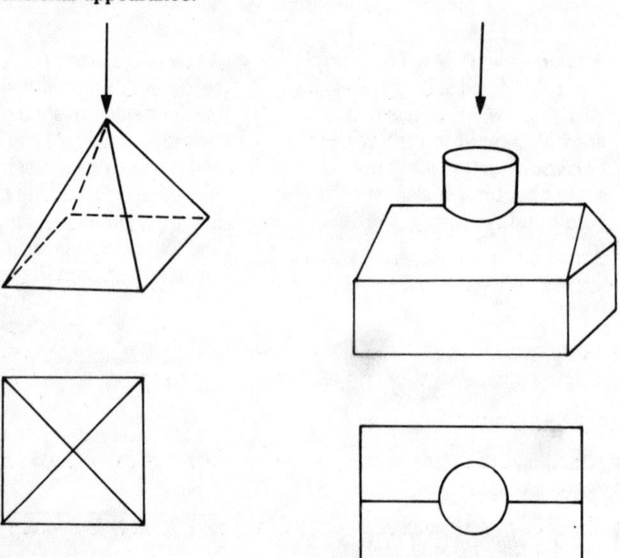

Fig P.19 Plans

POINT SYMMETRY

◀ Rotational symmetry ▶

POLYGON

A polygon is a plane figure with many straight sides. The names of the ones you ought to know are:

- **Triangle** 3 sides
- **Quadrilateral** 4 sides
- **Pentagon** 5 sides
- **Hexagon** 6 sides
- **Septagon** 7 sides
- **Octagon** 8 sides
- **Nonagon** 9 sides
- **Decagon** 10 sides

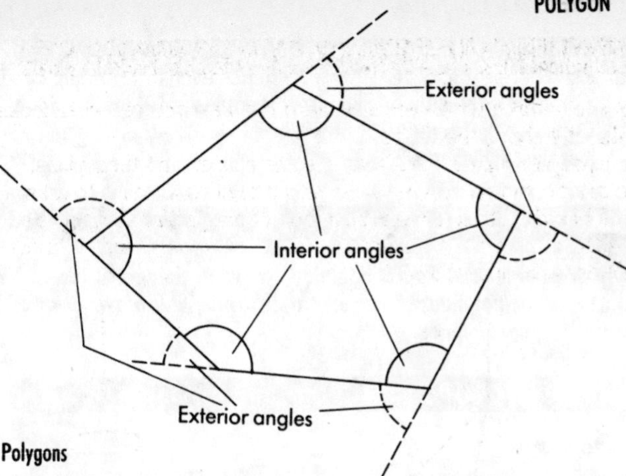

Fig P.20 Polygons

Polygons have two main types of angle. There are *interior angles* and *exterior angles* (outside), as shown in Figure P.20. A polygon will have as many exterior angles as interior angles, which will be the same as the number of sides of the polygon:

- Exterior angles: all the exterior angles of any polygon will add up to 360°.
- Interior angles: all the interior angles of a polygon add up to $180 \times (N-2)°$

Regular polygons

A *regular* polygon is one which has all its sides the same length and all its angles are the same.

- Exterior angle: the *exterior angle* of a regular N-sided polygon is found by dividing 360° by N.
- Interior angle: the *interior angle* of a regular N-sided polygon can be found by either subtracting the exterior angle away from 180°, or by using the formula $\frac{180\,(N-2)°}{N}$.
- Symmetry: an N-sided regular polygon will have N **Lines of symmetry** and **Rotational symmetry** of order N.
- **Exam Question**
  Figure P.21 shows a regular pentagon inscribed in a circle of centre O.
  a) Prove that the total of all of the interior angles of the pentagon (at A, B, C, D and E) is 540°. Explain your reasoning.
  b) Calculate the size of angle ADB.

  (NEA; H)

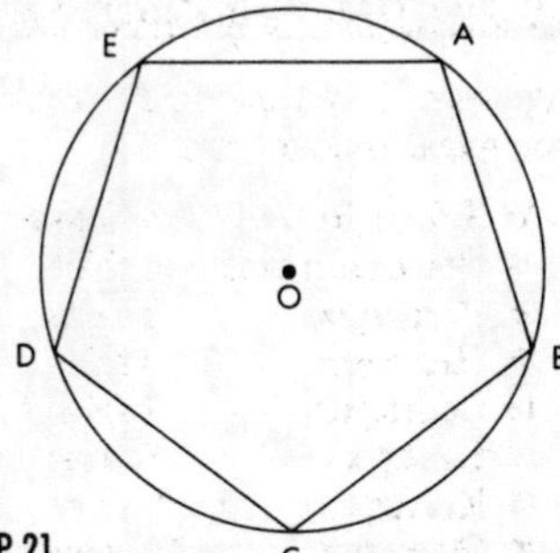

Fig P.21

■ **Solution**

a) Each exterior angle is $\frac{360}{5} = 72°$, hence each interior angle is 180 − 72 = 108. So, the sum of all 5 interior angles is 108 × 5 = 540°.

b) Consider the angle AOB; it will be $\frac{360}{5} = 72°$, then the angle ADB is half this, which is 36°.

POLYHEDRON/POLYHEDRA

A polyhedron is a solid figure bounded by plane polygonal faces. There are various types of polyhedra. For example, the *regular tetrahedron* has four equilateral triangular faces.

POSITION VECTORS

◀ Vectors ▶

POWER

◀ Exponent ▶

PREMIUM

A premium is an amount paid for an insurance contract. You can have annual, monthly or even weekly premiums. For example, you can see from the insurance table in Figure P.22 that the premiums vary for different ages and for the amount of insurance required.

12 MONTHLY REPAYMENTS
APR 23.6%

	Without Protection Insurance		*With Protection Insurance*		
Amount of loan £	*Monthly payment* £	*Total amount payable* £	*Premium* £	*Total amount payable* £	*Monthly payment* £
500	46.66	559.92	13.00	574.44	47.87
600	56.00	672.00	16.00	689.88	57.49
700	65.33	783.96	18.00	804.12	67.01
800	74.66	895.92	21.00	919.44	76.62
900	84.00	1008.00	24.00	1034.88	86.24
1000	93.33	1119.96	26.00	1149.00	95.75
1100	102.66	1231.92	29.00	1264.32	105.36
1200	112.00	1344.00	32.00	1379.76	114.98
1300	121.33	1455.96	34.00	1494.00	124.50
1400	130.66	1567.92	37.00	1609.32	134.11

Fig P.22 Premiums

PRIME

A number is prime when it has exactly two **factors** which are itself and the number one. The first few *prime numbers* are 2, 3, 5, 7, 11, 13 and 17; as you see they all have only two factors. It should be noted that the number one is *not* a prime number as it has only one factor.

Prime factor

The prime factors of an **integer** are the factors that are prime numbers. For example, the prime factors of 33 are 3 and 11, which for convenience in this sense we write as 3×11. The prime factors of 18 are $2 \times 3 \times 3$ (note how we put the 3 down twice so that the product of these factors gives the integer we start with). You could check for yourself that the prime factors of 48 are $2 \times 2 \times 2 \times 2 \times 3$ which we would shorten to $2^4 \times 3$.

PRISM

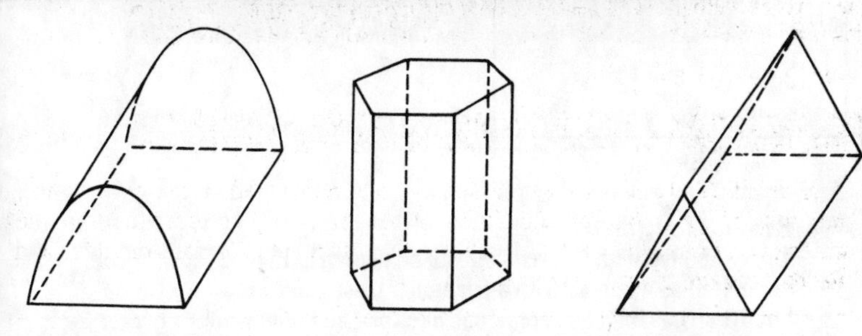

Fig P.23 Prisms

A prism is a three-dimensional shape with a regular cross section through its height or its length. All the shapes in Figure P.23 are prisms, since they are shapes you could 'slice' up in such a way that each cross section would be identical.

The volume of any prism is found by multiplying the area of the regular cross section by its length (or height if it is stood on its regular cross section).

For example, the volume of the prism in Figure P.24 is the area of the triangular end multiplied by the length, which is:

$\frac{1}{2} \times 5 \times 6 \times 8 = 120\text{cm}^3$.

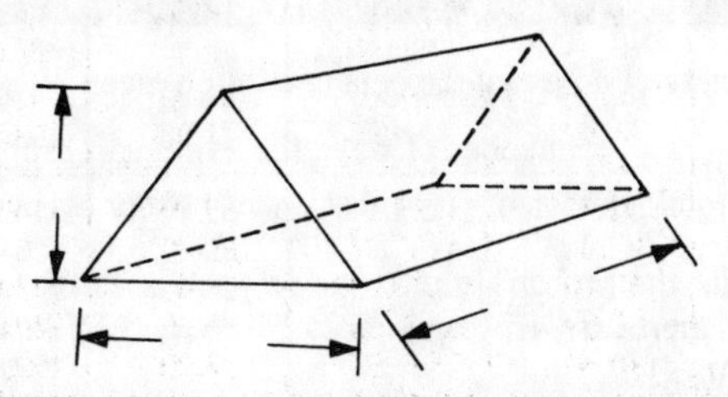

Fig P.24

Fig P.25

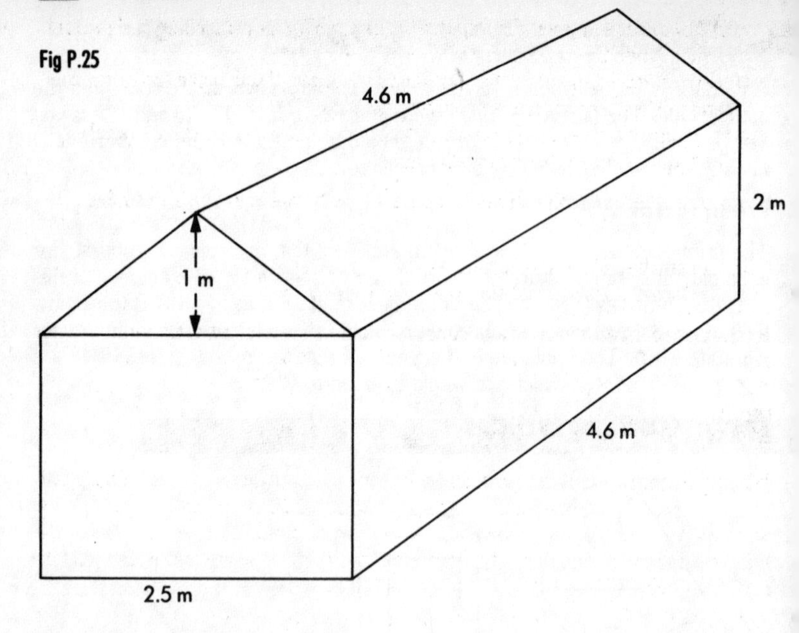

■ **Exam Question**

A greenhouse with the size and shape shown in Figure P.25 can be thought of as a cuboid with a triangular prism on top.
Mr. Khan has built a greenhouse like this, and now wants a heater for it. To choose the right heater, he needs to know the volume of his greenhouse.
Calculate the volume of the greenhouse.

(NEA; I)

■ **Solution**

The area of the end is a rectangle + a triangle, which is
$(2 \times 2.5) + (\frac{1}{2} \times 2.5 \times 1) = 6.25\text{m}^2$
So, the volume $= 6.25 \times 4.6 = 28.75\text{m}^3$

PROBABILITY

The probability of an event happening is often given by a *fraction*:

$$\frac{\text{the number of ways the event can happen}}{\text{the total number of ways that equally likely events can happen}}$$

For example, the probability of cutting a pack of cards and getting an ace is 4/52, since there are 4 possible aces to get and 52 equally likely cards to choose from.
There are three important probabilities to know:

- *No chance:* if an event is *impossible,* like rolling a normal dice and getting a nine, then the probability is 0.
- *Even chance:* an event that we say has an *even chance,* like tossing a coin and getting a head, has a probability of $\frac{1}{2}$.
- *Certainty:* if an event is *certain,* like rolling a dice and getting a number less than 7, then the probability is 1.

So you can see, *all* probabilities lie somewhere between 0 and 1 inclusive. Of course:

- the smaller a fraction is, the less likely the event
- the larger a fraction is, the more likely the event.

If the probability is given by a decimal, it can be changed into a fraction, e.g. $0.2 = \frac{2}{10}$.

COMBINED EVENTS

When we want to find the probability of a *combined event* – that is, where two or more events are happening – then we need to be clear about whether we want two events to happen *at the same time* or whether either event can happen, but *not* at the same time. These two situations are types that can be described as AND and OR.

AND

AND is the type where both events do happen at the same time. To find this combined probability we *multiply* the probabilities of each single event.
For example, what is the probability of rolling a normal dice twice and getting a three followed by an even number?
The probability of rolling a three is $\frac{1}{6}$
The probability of rolling an even number is $\frac{3}{6}$
Hence the combined probability is $\frac{1}{6} \times \frac{3}{6}$ which is $\frac{3}{36}$ which cancels down to $\frac{1}{12}$.

OR

OR is the type when either one event or the other can happen, but both cannot happen at the same time. To find this combined probability we *add* together their probabilities.
For example, to win a game, Terry had to cut a pack of cards and get a king or the Ace of Spades. What is the probability of this happening?
The probability of cutting a king is $\frac{4}{52}$
The probability of cutting an Ace of Spades is $\frac{1}{52}$
Hence the combined probability is $\frac{4}{52} + \frac{1}{52}$ which is $\frac{5}{52}$.

TREE DIAGRAMS

Tree diagrams are a very useful way of looking at *all* the possibilities of a given situation and they help us find various specific probabilities.
For example, two twins Helen and Neil both swim for their country in an international competition. The coach quotes the probabilities of them getting

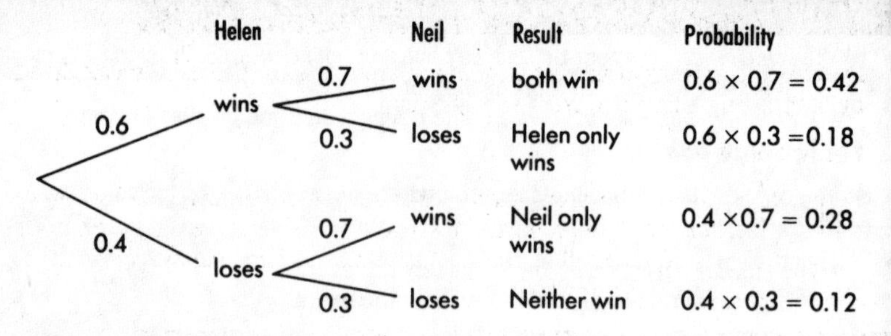

Fig P.26 Tree diagram

gold medals as Helen 0.6 and Neil 0.7. What are the chances of them both or either betting gold medals?

We can illustrate the possibilities with a tree diagram, as in Figure P.26. Notice how the individual probabilities have been put on the *branches* of the tree diagram, with each pair adding up to 1; this is because, for each pair, one of the branches *must* happen. Then to find the probabilities of the **Combined events** you just multiply along the branches, as shown.

EXPECTATION

The expectation of any event happening is the probability of that event multiplied by the number of times the 'trial' is done.

Suppose, for example, that the probability of a firm manufacturing a dud light bulb was 0.02. Then, if one day they produced 560 light bulbs, they would *expect* to have 0.02×560 dud ones, which is 11.2 (round off to 11).

- **Exam Question**

  A bag contains five discs, identical in every way except that one is white and the other four are red. Two boys, Alan and Ben, play a game whereby each takes it in turn to draw a disc at random from the bag without replacing it. The first one to draw the white disc is the winner. Given that Alan goes first, find the probability that:
  a) Alan wins at his first attempt,
  b) Ben wins at his first attempt,
  c) Ben wins.

  (NEA; H)

- **Solution**
- a) $\frac{1}{5}$
- b) The probability is that Alan loses first and then Ben wins, which is $\frac{1}{5} \times \frac{1}{4} = \frac{1}{5}$
- c) For Ben to win, there are only two possibilities: Alan lose, Ben win, **or** Alan lose, Ben lose, Alan lose, Ben win.
  These chances will be $\frac{1}{5}$ **or** $\frac{4}{5} \times \frac{3}{4} \times \frac{2}{3} \times \frac{1}{2}$

  $$= \frac{1}{5} + \frac{1}{5} = \frac{2}{5}$$

PROFIT

Profit is the amount of money gained by a transaction. It is usually calculated by subtracting cost price from selling price.

Percentage profit

Percentage profit is the profit divided by the cost price multiplied by 100 to turn the fraction into a percentage. For example, if a man bought a plant for 25p and then sold it for 40p, the profit is 15p, and his *percentage* profit is calculated as $^{15}/_{25} \times 100$ which is 60%.

PROPORTION

Proportion is the relation of one number with another to form a **ratio**. There are three main types of proportion; direct, inverse and joint.

DIRECT PROPORTION

Direct proportion is when there is a simple *multiplying* connection between two things. For example, if a spoon weighs 25g, then the weight of any number of the same sized spoons is found by multiplying the number of spoons by 25g. This is because the weight of the spoons is in *direct proportion* to their number.

This proportion, or **variation** as it is often called, is often related to the square or the cube of something. For example, the *volume of a sphere* is in direct proportion to *the cube of its radius*.

The alternative way of saying 'is directly proportional to' include the following:

- the weight *varies directly with* the number
- the weight α the number
- the weight = K × (the number).

The last two are mathematically the most convenient to use and you could use this shorthand a lot. The k is a constant and is called the *constant of proportionality*.

For example, the mass of a ball bearing varies directly with the cube of the radius. A ball bearing with a radius of 3cm has a mass of 450g. Find the mass of a similar ball bearing with a radius of 2cm.

Since mass α radius^3, then mass $= K \times R^3$

so when mass = 450g and $R = 3$, then $450 = K \times 27$; hence $K = 16.666$.

Therefore, when $R = 2$, then mass $= 16.666 \times 8 = 133.3$g.

INVERSE PROPORTION

Inverse proportion is when there is a *dividing* connection between two things, so that as one increases the other decreases. For example, the more men involved in digging a hole, the quicker it will get dug (in theory anyway).

For example, the time taken to construct a building varies inversely with the square root of the number of people constructing. If a building is constructed by 15 men in 28 days, how long will it take 21 people?
Since $T \alpha 1/\sqrt{N}$, then $T = K/\sqrt{N}$
When $T = 28$, $N = 15$; hence $28 = K/\sqrt{15}$ and $K = 28 \times \sqrt{15}$
So when $N = 21$, $T = \dfrac{28 \times \sqrt{15}}{\sqrt{21}} = 23.66$ (call it 24 days).

JOINT PROPORTION

Joint proportion is where three (or more) things vary with each other in *combinations* of direct and/or inverse proportion.

For example, the breaking point (BP) of a wooden bridge varies directly with the width of the bridge but inversely with its length. A wooden bridge that was 1 metre wide and 5 metres long had a BP of 160 kg. What is the BP of a bridge that is 2 metres wide and 8 metres long?
Since BP α W/L, then BP = K × W/L
when W = 1 and L = 5, then BP = 160
hence 160 = K × 1/5, so K = 160 × 5 = 800
So, when W = 2 and L = 8 then BP = 800 × 2/8, which is 200 kg.

- **Exam Question**
  A mathematician looking for Christmas presents noticed that three types of wine glass in a certain store (Fig P.27), were all similar in shape and that:

$$\frac{\text{radius of base of type A}}{\text{radius of base of type B}} = \frac{\text{radius of base of type B}}{\text{radius of base of type C}} = \frac{2.}{3}$$

Fig P.27

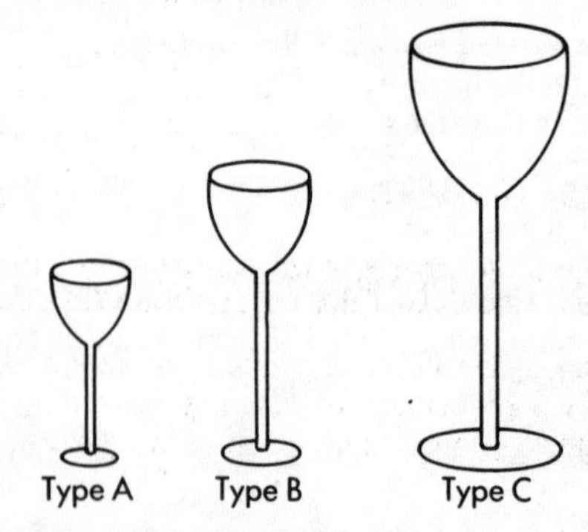

a) Calculate

i) $\dfrac{\text{perimeter of bowl of glass type C,}}{\text{perimeter of bowl of glass type B}}$

ii) $\dfrac{\text{radius of base of glass type C,}}{\text{radius of base of glass type A}}$

iii) $\dfrac{\text{volume of bowl of glass type A.}}{\text{volume of bowl of glass type B}}$

b) If the prices of the wine glasses were in direct proportion to the volume of bowl, and glass type B cost £3.51, what would be the price of glass type A?

(NEA; H)

■ **Solution**

a) i) $\frac{3}{2}$; ii) $(\frac{3}{2})^2 = \frac{9}{4}$; iii) $(\frac{2}{3})^3 = \frac{8}{27}$

b) $£3.51 \times \frac{8}{27} = £1.04$

PYRAMID

A pyramid can have any shape for its base. But from each point on the perimeter of the base there is a straight line going up to the top and all these lines will meet at one point, called the *vertex*. Figure P.28 shows a *square-based* pyramid (often the name of the pyramid will reflect the shape of the base).

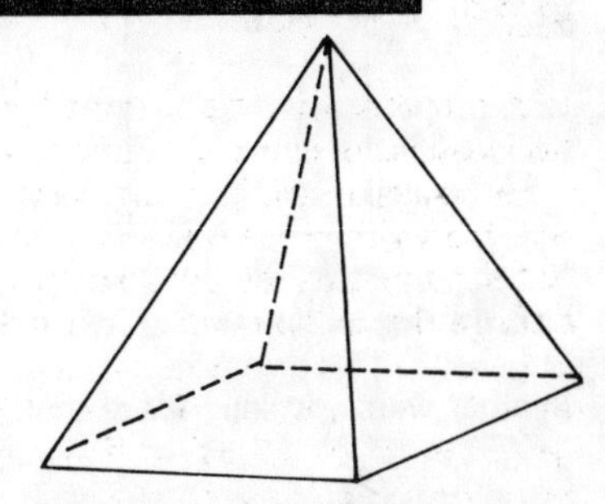

Fig P.28 Square-based pyramid

A pyramid with a *circular base* however is called a **cone** (Fig P.29). When the vertex of a pyramid is vertically above the centre of the base, the correct name is a 'right pyramid'.

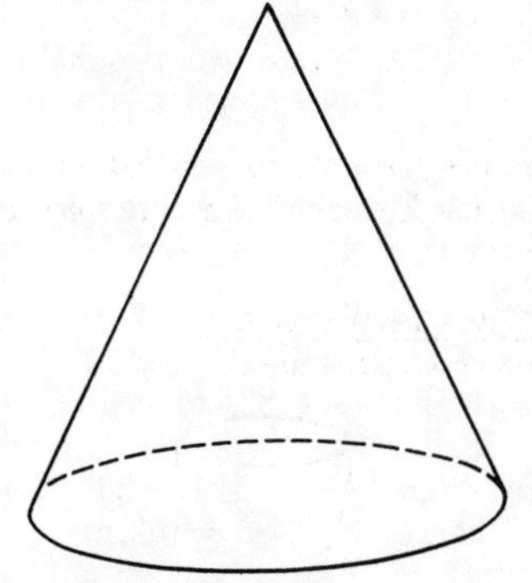

Fig P.29 Cone (circular-based pyramid)

The *volume* of any pyramid is found by multiplying its base area by one third of its height:

$$\text{volume} = \text{base area} \times \tfrac{1}{3}\,\text{height}$$

PYTHAGORAS

Pythagoras was a Greek mathematician who later lived in South Italy in the latter half of the sixth century BC. He organised many groups of scholars to probe into geometry, philosophy, religion and politics.

PYTHAGORAS' THEOREM

Many researchers now believe that it was *not* Pythagoras who first *discovered* this theorem, although it is quite likely that Pythagoras did, in his own time, find it out for himself. The theorem states that:

'In a right-angled triangle, the sum of the squares of the two smaller sides is equal to the square of the longest side.'

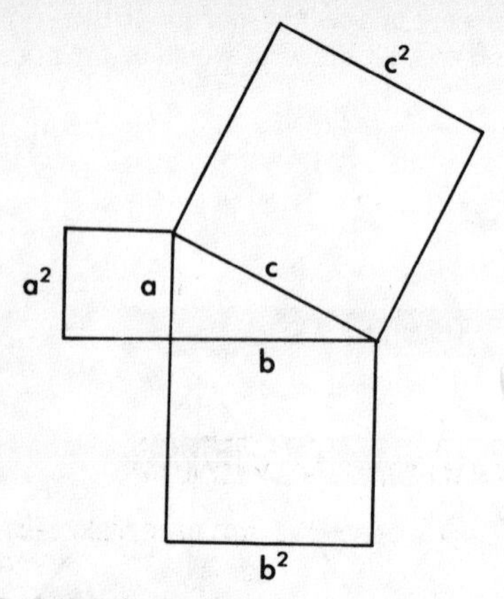

Fig P.30 Pythagoras' theorem

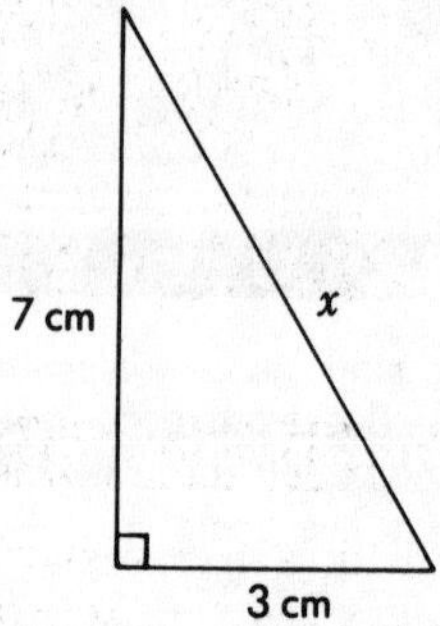

Fig P.31

In other words, looking at the triangle in Figure P.30, with sides a, b and c:

$a^2 + b^2 = c^2$

You need to be able to use the formula to find missing sides in right-angled triangles when you know the other two sides.

For example, in Figure P.31, to find x, use Pythagoras' theorem to state $x^2 = 3^2 + 7^2 = 9 + 49 = 58$; hence $x = \sqrt{58} = 7.6$cm

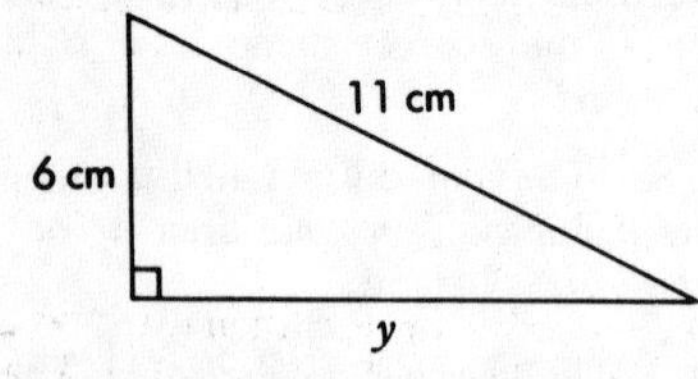

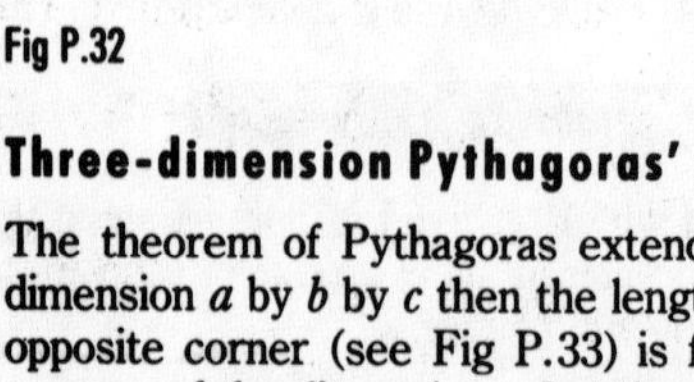

Fig P.32

In Figure P.32, to find y use Pythagoras' theorem to state $y^2 = 11^2 - 6^2 = 121 - 36 = 85$; hence $y = \sqrt{85} = 9.2$cm

Three-dimension Pythagoras' theorem

The theorem of Pythagoras extends into *three dimensions*. In a cuboid of dimension a by b by c then the length of the diagonal from one corner to the opposite corner (see Fig P.33) is found by square rooting the sum of the squares of the dimensions. In other words:

length of diagonal $= \sqrt{(a^2 + b^2 + c^2)}$

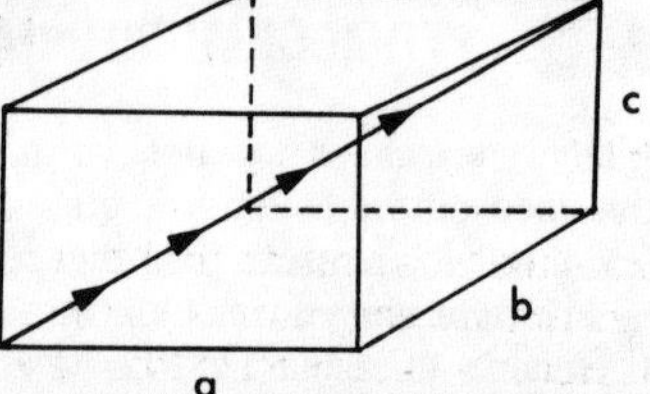

Fig P.33 Three-dimensional Pythagoras' theorem

QUADRATIC

A quadratic is an expression that involves powers of two, but none higher than 2; for example $x^2 + 5x - 9$

Factorisation of quadratic equations

◀ Equations ▶ (in particular *quadratic equations*).

Difference of two squares

If you expand $(x + y)(x - y)$ in the normal way you get the quadratic $x^2 - y^2$. This is called the *difference of two squares*; it is very useful to recognise when this is present, especially when we wish to **factorise**. The situation is recognisable as being the subtraction of two expressions that can be *square rooted* quite easily.

For example, to solve $9x^2 - 16 = 0$, we recognise a difference of two squares.

This gives us $(3x + 4)(3x - 4) = 0$

hence $x = -4/3$ and $4/3$

Graphs of quadratic equations

If the equation is of a quadratic nature, i.e. $y = ax^2 + bx + c$, then it will be a ∪- shaped curve if a is positive, and a ∩-shaped curve if a is negative. The value of c is the y-axis intercept.

For example, Figure Q.1 shows a sketch of the graph of $y = 3x^2 + 2x - 1$.

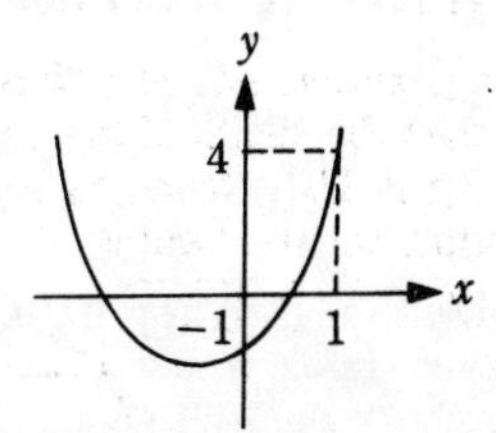

Fig Q.1

Since the 'a' is positive we know it's a ∪-shape, since the 'c' is -1 we know it cuts through the y-axis at $y = -1$, and when $x = 1$, $y = 3 + 2 - 1$ which is 4. Hence a sketch could look like that shown.

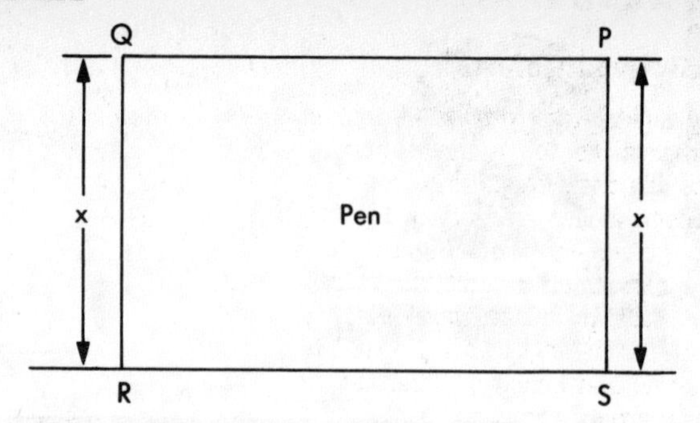

Fig Q.2

- **Exam Question**

A farmer has 16m of fencing which he is going to use to make a rectangular pen *PQRS* in which to keep chickens.
He has been advised that he should allow 1 m^2 for each chicken. The pen is to be built using a barn wall as one side, as shown in Figure Q.2. He wants to design the pen in such a way that he can keep the maximum number of chickens in it.
Let the length of *PS* be x metres.

a) Write down, in terms of x, an expression for the length of the side *PQ*.

b) Show that the area, *A* square metres, of the pen is given by $A = 2x\ (8-x)$.

c) Complete the table below, which gives the value of *A* for different possible values of x.

x	0	1	2	3	3.5	4	4.5	5	6	7	8
A		14	24	30	31.5			30	24	14	0

d) Draw the graph of *A* against x. Use your graph to find the maximum number of chickens that the farmer may keep in the pen and calculate the dimensions of the pen that he should use.

(NEA; H)

- **Solution**

a) PQ = $16-2x$

b) Area = $x\ (16-2x)$
$= 16x - 2x^2 = 2x(8-x)$

c)

x	0	4	4.5
Area	0	32	31.5

d) The graph will be a ∩-shape from (0,0) to (8,0).
maximum chickens (area) = 32
dimension of pen = 4×8m.

QUADRILATERAL

A quadrilateral is a four-sided plane figure. Its inside angles add up to 360°.

Cyclic quadrilateral

Any quadrilateral drawn so that its four vertices touch the circumference of a circle is said to be *cyclic*. Its opposite angles will add up to 180°. For example, in Figure Q.3 $(a + c) = (b+d) = 180°$. It is also true that any quadrilateral that has opposite side angles adding up to 180° is cyclic and hence a circle can be drawn around the vertices.

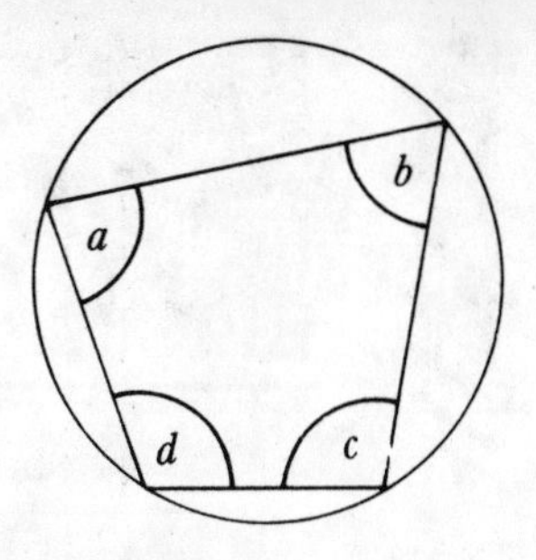

Fig Q.3 Cyclic quadrilateral

QUARTILE

Quartiles are found by dividing a cumulative frequency (c.f.) into four quarters. The points on the c.f. that give us the quartiles can be found by dividing the total frequency into four equal groups.

As you see on the diagram of a c.f. (Fig Q.4), we actually have three quartiles. The lowest line gives us the *lower quartile*; the middle line gives us the *median*; and the highest line gives us the *upper quartile*.

In general to find the quartiles of a distribution of a frequency N:

- the *lower quartile* is found at $(N + 1)/4$ on the c.f.;
- the *median* is found at $(N + 1)/2$ on the c.f.;
- the *upper quartile* is found at $3 \times (N + 1)/4$ on the c.f.

Interquartile range

The interquartile range is the numeric difference between the upper quartile and the lower quartile. For example, in Figure Q.4, the upper quartile is 16.6 and the lower quartile is 11.2, so the interquartile range is $16.6 - 11.2 = 5.4$

Semi-interquartile range

This is exactly what it says: half of the interquartile range. In our example, the semi-interquartile range is $\frac{5.4}{2} = 2.7$

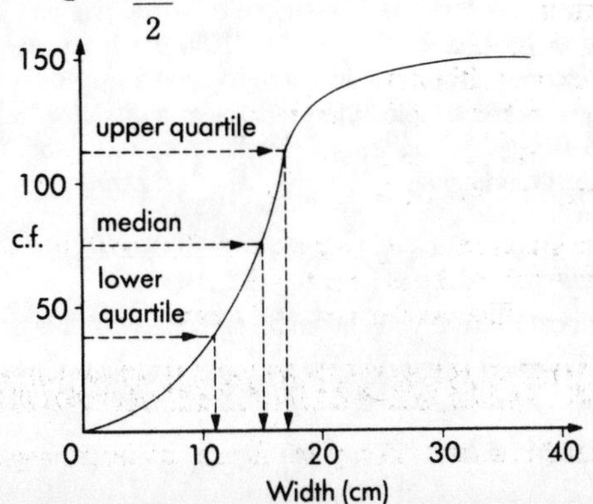

Fig Q.4 Quartile

RADIUS

The radius of a circle is the distance from the centre of the circle to the **circumference**. It is exactly half of the **diameter** of a circle.

RANGE

Range has two meanings in mathematics:

1. We have seen that the **domain** is the set of numbers that the given **function** is applied to. The range of a *function* is the set of numbers that the function takes those numbers to. Each number or element starts in the domain and its 'image' under the function ends up in the range.
2. The *range* of a *set of data* is simply the difference between the highest and the lowest value. For example, in a set of data about heights of pupils, Sean was the tallest with 195cm and Simon was the smallest with 98cm. The range is 195–98, which gives us 97.

RATE

A rate is a fixed ratio between two things. The idea of 'rate' is used quite a lot; e.g. speed, which is the rate of distance travelled per unit of time, or costs of hiring items, say a boat at the rate of £1.50 per hour.

The *gradient* of a graph will usually give you a rate of some sort connected with the units. For example, the gradient of a distance/time graph gives you the *rate of change of distance*, which is **speed**; and the gradient of a velocity/time graph gives you the *rate of change of velocity*, which is **acceleration**.

Speed

This is an important rate you should be familiar with; it is the rate of change of distance. It is useful to know that:

$$\text{speed} = \frac{\text{distance travelled}}{\text{time taken}}$$

Exchange rates

◀ Exchange rates ▶

■ **Exam Questions**

1 Margaret and David want to find out how much petrol their car uses in miles per gallon. They fill the car with petrol when the reading on the milometer is

0 2 8 3 4 0

After a few days they stop at a garage selling petrol at £1.70 per gallon. They fill the car up again and the cost of the petrol is £13.60.

By this time the milometer reading is

0 2 8 6 2 0

How many miles per gallon does the car do over this period?

(WJEC; I)

2 According to the Guinness Book of Records, the steepest temperature rise ever recorded occurred in South Dakota on 22 January 1943 when the temperature rose from −20°C at 7.30 a.m. to 7.2°C at 7.32 a.m.

a) By how much did the temperature rise?
b) What was the average rate at which the temperature was rising in °C/sec?

(NEA; I)

■ **Solutions**

1 The distance covered is 280 miles, the petrol used is 8 gallons, hence the miles per gallon will be 280÷8 which is 35 mpg.

2 a) 7.2°C − − 20°C = 27.2°C
b) 27.2°C per 2 minutes, i.e. 120 seconds.
hence 27.2 ÷ 120 = 0.227°C/sec.

RATIO

Ratio is a comparison between two (or more) amounts, often written with 'to' or a colon (:). Many mathematical problems are sorted out by the formal use of ratio.

A ratio such as 12:8 can be simplified (or cancelled down) in a similar way as fractions can be, by dividing both sides by the same amount. Here 12:8 can be simplified to 3:2

For example, a good blend of tea can be made from Gunpowder tea and Lapsang in the ratio of 5:3. How much of each tea is there in a 500g packet?

Add together the parts of the ratio to get 8, and then divide this into 500g to get 62.5g. Multiply this amount by the two individual parts of the ratio to give:

Gunpowder tea . . . 5 × 62.5g = 312.5g and
Lapsang tea . . . 3 × 62.5g = 187.5g

■ Exam Questions

Figure R.1 shows a recipe for making 16 biscuits.

a) How many biscuits can be made using 12 oz of plain flour if there is enough of all the other ingredients available?

b) How much plain flour is needed for 24 biscuit?

(NEA; L)

LEMON BUTTER BISCUITS
6 oz Butter
2 oz Icing sugar
½ teaspoon Vanilla essence
1 Lemon, finely grated
4 oz Plain flour
2 oz Cornflour
1 Egg, beaten

Fig R.1

■ Solutions

a) $12 \div 4 = 3$; hence three lots of 16 can be made, which is 48.

b) 4 oz are used for 16 biscuits
so 1 oz is used for 4 biscuits
hence 6 oz will be needed for 24 biscuits.

RATIONAL

A rational number is one which can be expressed as a vulgar fraction, i.e. as a/b, where both a and b are integers. A rational number will always be either a **terminating decimal** or a **recurring decimal**.

Irrational

Any number that is *not* rational is irrational. The two most quoted irrational numbers are π and $\sqrt{2}$.

REAL NUMBER

A real number is one that is either **rational** or **irrational**. All the possible numbers that you can come across in a GCSE course are real numbers. To find out about numbers that are *not* real then you need to study A level mathematics, where for instance you can find out about *imaginary* numbers.

RECIPROCAL

The reciprocal of a number is the result of dividing that number into 1. For example the reciprocal of 5 is $\frac{1}{5}$.

The reciprocal of any **vulgar fraction** is that vulgar fraction turned upside down; for example the reciprocal of $\frac{3}{4}$ is $\frac{4}{3}$.

On many calculators there is a button that calculates reciprocals for you, you will recognise it as [1/x]

Reciprocal equations

◄ Equations, Graphs ►

RECTANGLE

A rectangle is a four-sided, plane shape where the opposite sides are parallel as well as being of the same length. All the angles of a rectangle are right angles. It has two **lines of symmetry**, namely each line bisector, and it has **rotational symmetry** of order two.

The area of a rectangle is found by multiplying the base length by the height.

RECTANGULAR BLOCK

◀ Cuboid ▶

RECURRING

◀ Decimal ▶

REFLECTION

A mathematical reflection is the mirror image of a shape drawn on the opposite side of a mirror line, such that each line drawn from one point to its reflection is perpendicular to the mirror line.

When asked to draw a reflection of a shape in a given line, then there are two ways of doing it:

1 If you were asked to reflect the triangle ABC (Figure R.2) in the line XY, you could trace the figure ABC as well as the mirror line XY. Then, flip the tracing over, so that the figure ABC appears *under* the line XY with the traced line XY exactly on top of the original XY (make sure the X is on top of X and the Y on top of Y). Then, with a pencil, press down on the vertices of the triangle to make a 'dint' in the paper underneath. Take the tracing paper away and join up the dots; this will give you the reflection. Do this carefully and it is a very good way.

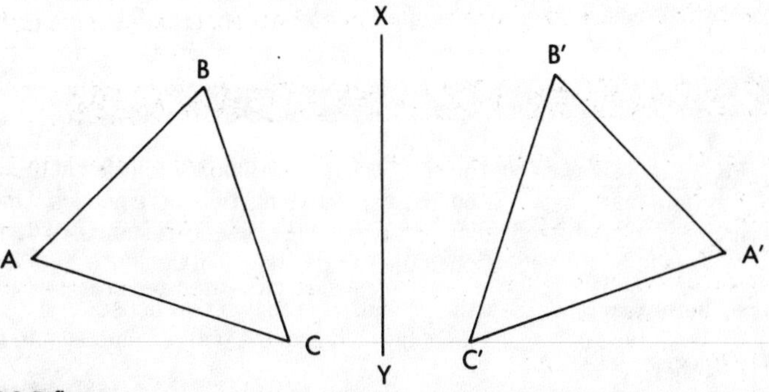

Fig R.2 Reflection

2 The other way is to draw faint lines from each vertex of the triangle perpendicular to the line XY and to extend each line the other side of XY. Make line the same distance either side of the mirror line, and the end points of the lines can be joined up to give you the reflected shape. If this is done on squared paper, as many in an examination are, then this method is the most accurate.

When finished, the mirror line should be a **line of symmetry** between the two sides.

The most common mistake made in reflections is to draw the same shape on the opposite side of the mirror line but either too near or too far away, and to get the 'line of symmetry' wrong, so that the reflection ends up also being translated.

■ Exam Question

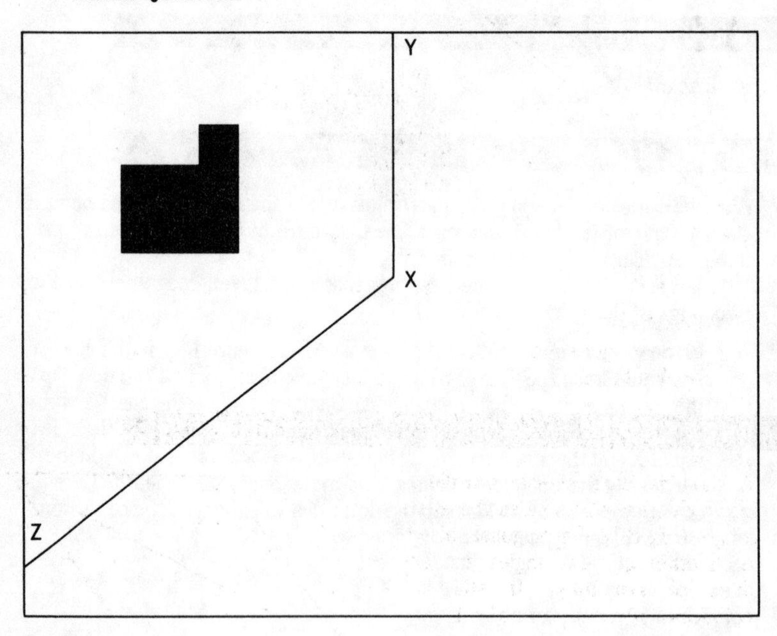

Fig R.3

Figure R.3 shows a plan of a plot of land with a house (shaded in the diagram) built on it. The builder wants to build two more houses. One house is to be the reflection of the existing house in the fence XY and the other is to be the reflection of the existing house in the fence XZ.

a) In the diagram, show the positions of the two new houses.
b) What type of transformation would map the two reflections onto each other?

(NEA; I)

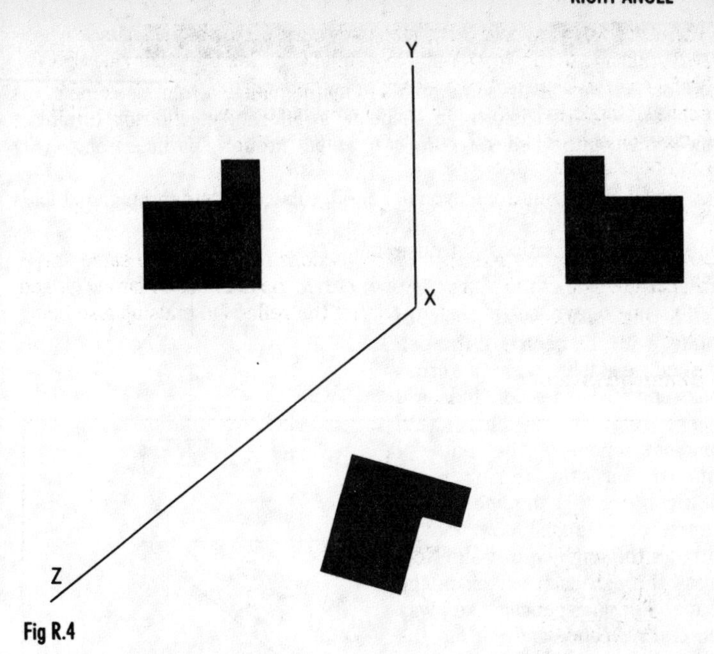

Fig R.4

■ **Solution**

a) See Figure R.4.
b) A rotation.

RHOMBUS

A rhombus is a special **parallelogram** that has all its sides the same length (Fig R.5). Its diagonals bisect each other at right angles and are **lines of symmetry**. It still has **rotational symmetry** of order two.

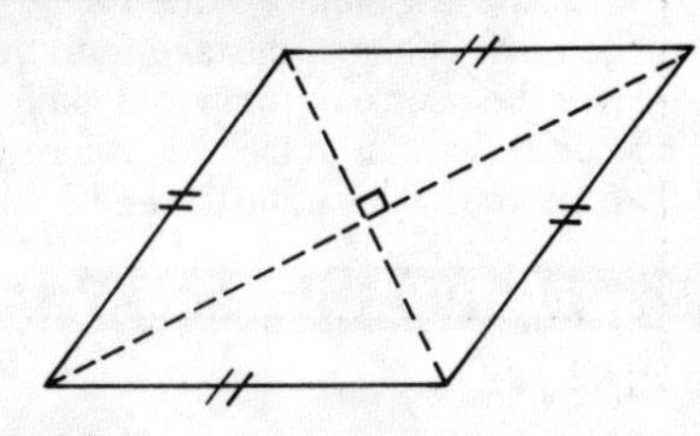

Fig R.5

RIGHT ANGLE

A right angle is a quarter turn, which is 90 degrees.

Right-angled triangles

The solution of right-angled triangles, that is finding lengths and angles, can be done by **trigonometry** and **Pythagoras**.

ROTATION

A **rotation** is a turn about some point. In mathematics, a rotation is meant as the process which transforms a shape to a different position by turning it around some point, called the *centre of rotation*, through a particular angle (Fig R.6).

For the GCSE examinations you need to be able to rotate through multiples of 90°.

One way to do rotations is to use tracing paper:

> Trace the shape you wish to rotate, with the centre of rotation marked suitably with a cross indicating the lines of the grid you're rotating on. Then, with your pencil-point on the centre of rotation, turn the tracing paper until you see that the lines at the centre of rotation have moved through the angle you wish. Now press through each vertex of the shape with your pencil; take away the tracing paper and join up the dots you have just made.

The shape produced in this way should be the same size as the original, but in a different position.

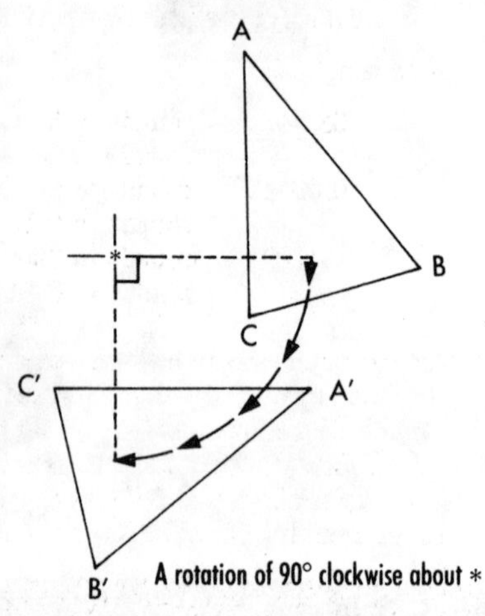

Fig R.6 Rotation

When you have practised quite a few rotations you should be able to draw them without having to trace each time but simply by thinking through where the shape is going to rotate to.

The most common mistake is to get the wrong centre of rotation. In most cases the centre of rotation is likely to be the origin, but not neccessarily so. So do be careful to look for where the centre of rotation actually is, and not where you might want to see it!

ROTATIONAL SYMMETRY

◀ Symmetry ▶

ROUNDING

Rounding off is the process by which we make approximate answers to a problem. It helps in presenting an answer to a sensible degree of accuracy. It is essential that you can round off properly, as it is a situation that will crop up in your examinations quite often.

To round off to a specific **decimal place** or **significant figure**, the rules are basically the same:

- Decide where you need to round off to; perhaps one decimal place or two significant figures or the nearest ten. If in doubt, use *one more significant figure* than has been used in the question.
- Then look at the **next** digit:

  if it's less than a 5, you round down.
  if it's 5 or bigger, you round up.

For example:

32.547 . . .	rounds to 32.5	to one decimal place
	rounds to 33	to the nearest whole number
0.0964 . . .	rounds to 0.1	to one significant figure
	rounds to 0.10	to two decimal places
34729 . . .	rounds to 35000	to the nearest thousand
	rounds to 30000	to one significant figure

The most common error is one which some calculators do, namely to cut short the final answer; in other words to just chop off the bits beyond what's wanted, instead of rounding off.

Also some candidates make problems for themselves by rounding off too soon; try not to round off until the *final answer* in a problem. Otherwise the earlier rounding off will make your final answer inaccurate.

ROUTE MATRICES

◀ Matrix ▶

SALARIES

Salaries are the amounts of money people earn in a year. They are usually paid in either 12 monthly payments throughout the year, or 13 payments every four weeks. If an examination question refers to someone being paid monthly, you should assume that it is *calendar months* they are being paid (that is 12 payments per year) and not *lunar months* (that is 13 payments a year). If it is the lunar month payment, then the question will say so.

SCALE

USING SCALES

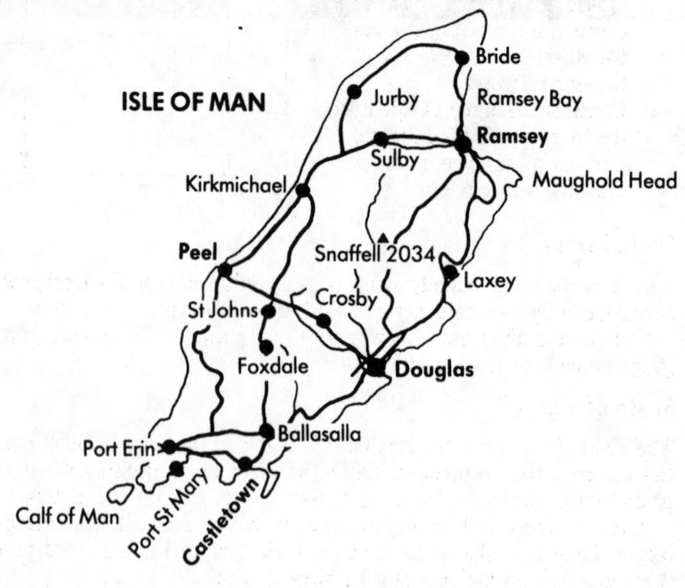

Fig S.1 Scale

Scale of a map

Maps have scales which are usually written down as a **ratio**; for example 1:1000 would mean that every 1cm on the map would represent 1000cm in reality (which is 1 metre).

For example, the scale on the map in Figure S.1 is 1:800,000.

Scale drawings

In the same way that a scale on a map is written as a **ratio**, we can use the same principle for buildings, say designing a house.

For example, on a *scale drawing* a scale of 1:100 would mean that every 1cm on the house plan would represent 100cm in the house.

- **Exam Questions**

  Mr. and Mrs. Johnston decided to join two rooms by removing part of the wall to form an archway. They sent a rough sketch to the builder who made an accurate scale drawing. Using the information in the sketch, construct a scale diagram using the scale 1:20.

  (NISEC; H)

- **Solutions**

  Look at Figure S.2. You should convert the given units to cm, then divide each by 20 to find the lengths to use.

  Construct each right angle from the base line, find the centre, then use a pair of compasses to draw the semi circle.

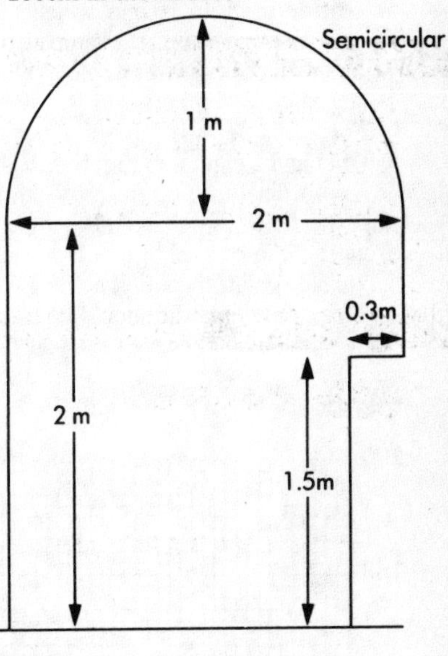

Fig S.2

Scale factor

A scale factor is the multiplying factor of an **enlargement**. Each original length is multiplied by the scale factor to find the new length.

Models also often have scale factors; for example the model railways 00 gauge is made on a scale of 1:100.

Scale of a graph

The scale of a graph is the relationship between the squares on the axes and the numbers they represent. Often, in examination questions, the scale is given to you; for example on the horizontal axis use a scale of 1cm to 20g.

If you are faced with having to *choose* your own scale, then remember, the bigger the scale – the more accurate is the graph. But also remember that there is a limit on the size of your graph paper.

You need to look at the *largest number* that you need to put onto each axis and then to see how best this fits in with the squares available on the paper. Take care to choose a scale that will help you to easily work out the position of the in-between numbers. For example, a scale going up in 3s, 4s or 7s etc., is no good since you cannot easily determine the numbers in-between. Your scale should go up in 2s, 5s, 10s, 20s, etc. When you have decided on the scale of your graph then do remember to fully *label* the axes; for example, time (seconds), distance (metres), or price (£), etc. Do not forget the *units* of the labels; these are important, you could lose marks if they are not there.

SCATTER DIAGRAMS

A scatter diagram is one which plots quite a few points representing *two* factors or *variables*. For example, people's different weight with their height; or age with their pocket money. The scatter diagram will help you see if there is any *relationship* between the two factors.

For example, we have scattered the information about some children's heights and weights on Figure S.3. Each child is represented by a blob which is a co-ordinate found from (weight, height). From the chart we can deduce that the taller the child, the heavier the child. But further than that we can, from the **line of best fit** that has been drawn, estimate the 'normal weight' for any child, given their height.

This line of best fit should be a straight line, unless the data is such that it shows an obvious curve that you can follow.

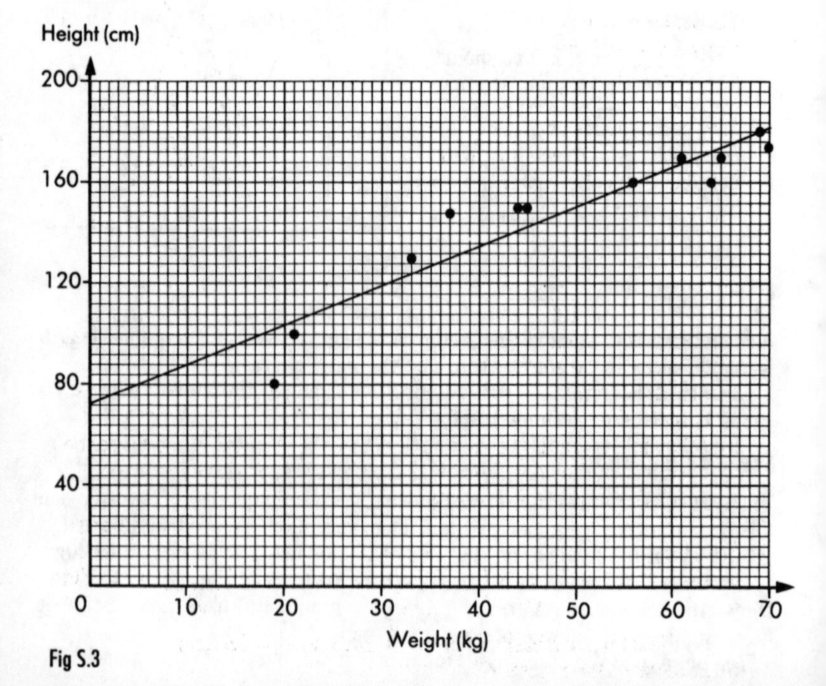

Fig S.3

Name	*height*	*weight*
James	160 cm	64 kg
John	150	45
Joseph	80	19
Paul	150	44
Michael	145	37
Jenny	100	21

Name	*height*	*weight*
Robert	160	56
Helen	165	61
Neil	165	65
Kirsty	130	33
Gary	180	69
Mark	175	70

SECTOR

A sector is part of a circle bounded by two radiuses (radii) and the arc between them, as in Figure S.4.

The *area* of a sector where the radius is r and the angle subtended at the centre of the circle is x, is given by: Area $= \frac{x}{360} \pi r^2$

The *arc length* of the sector is given by; arc length $= \frac{x}{360} \pi D$, where D is the diameter of the circle.

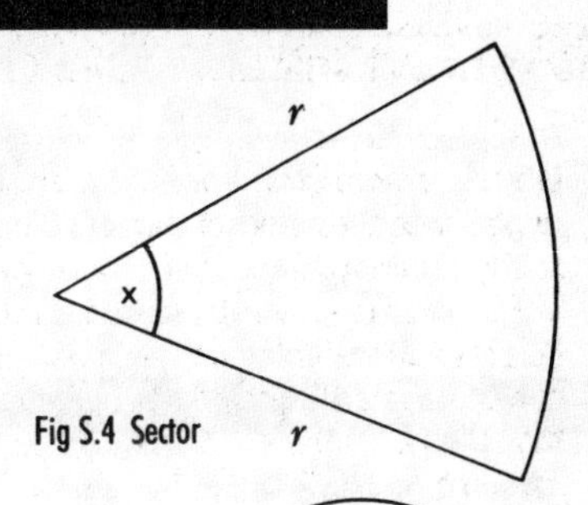

Fig S.4 Sector

■ **Exam Questions**

Figure S.5 represents a keyhole. The perimeter of the keyhole consists of two straight lines and two arcs of circles, one of radius 7 cm and the other of radius 2 cm. Both circles have centre O.

A formula for the length of an arc of a circle is given by

arc length $= \frac{\theta \times \pi \times r}{180}$

where r is the radius of the circle and $\theta°$ is the angle at the centre of the circle.

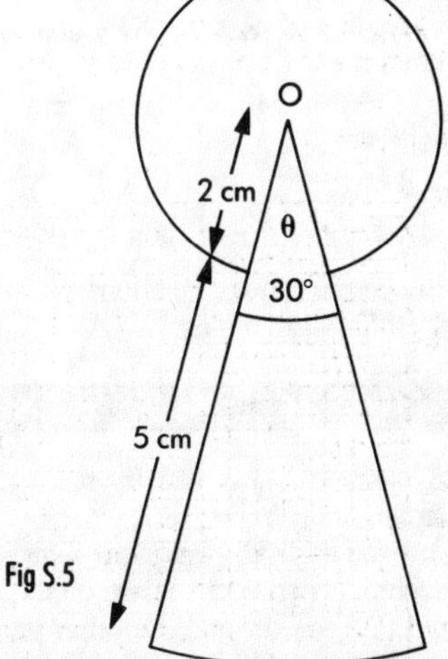

Fig S.5

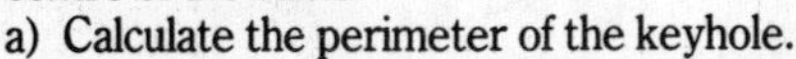

a) Calculate the perimeter of the keyhole.

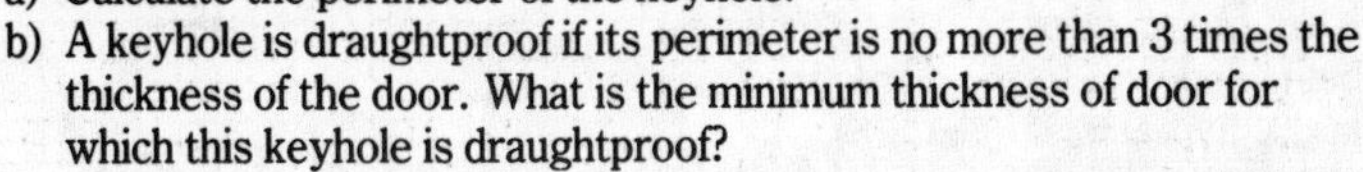

b) A keyhole is draughtproof if its perimeter is no more than 3 times the thickness of the door. What is the minimum thickness of door for which this keyhole is draughtproof?

(NEA; I)

■ **Solution**

a) Bottom arc $= \frac{30 \times \pi \times 7}{180} = 3.7$cm   Top arc $= \frac{330 \times \pi \times 2}{180} = 11.5$cm

So the total perimeter will be $3.7 + 11.5 + 10 = 25.2$cm

b) $25.2\text{cm} \div 3 = 8.4$cm.

SEGMENT

When a circle has a chord drawn in it, as in Figure S.6, then the chord divides the circle into two segments. The larger one is called the *major segment*; the smaller one is called the *minor segment*.

Angles in a segment

◀ Angles ▶

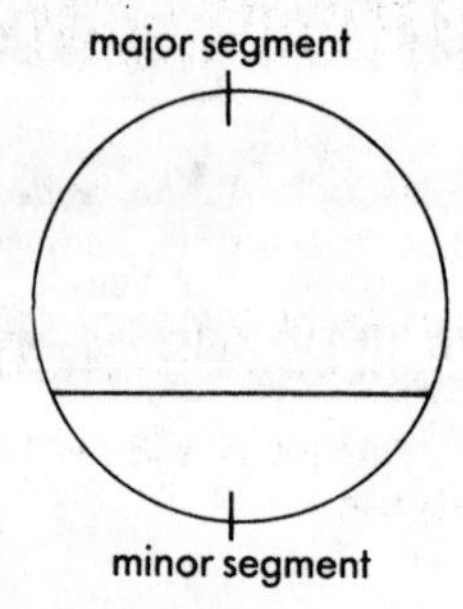

Fig S.6 Segment

SELF INVERSE

A function has a self inverse when the same function will return each number in an **image** back to the original number in the **domain**. Try the following for yourself and see that they are *all* self inverses:

If $f(x) = \frac{1}{x}$ then $f^{-1}(x) = \frac{1}{x}$

If $f(x) = \frac{24}{x}$ then $f^{-1}(x) = \frac{24}{x}$

If $f(x) = 10 - x$ then $f^{-1}(x) = 10 - x$.

Hence the type of functions $f{:}x \rightarrow \frac{A}{x}$ and $f{:}x \rightarrow A - x$ when A is a **real number** are *all* self inverses.

SEMICIRCLE

A semicircle is exactly half of the circle, as in Figure S.7.

A *triangle* drawn in a semicircle such that two vertices are on the end of the diameter and the other vertex is somewhere on the arc, is a *right-angled triangle*, the right angle being opposite to the diameter which is the hypotenuse (Fig S.8).

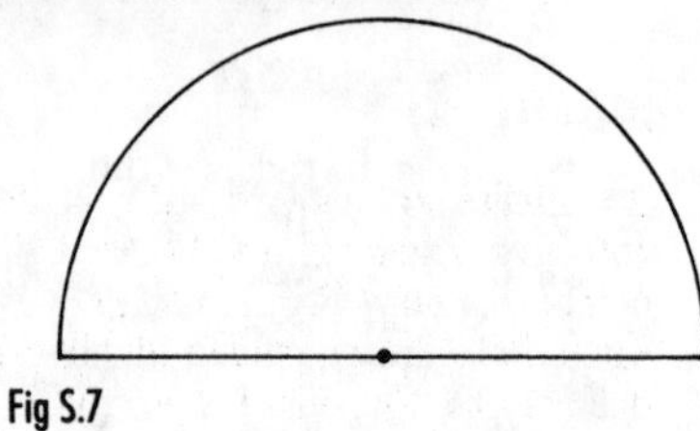

Fig S.7

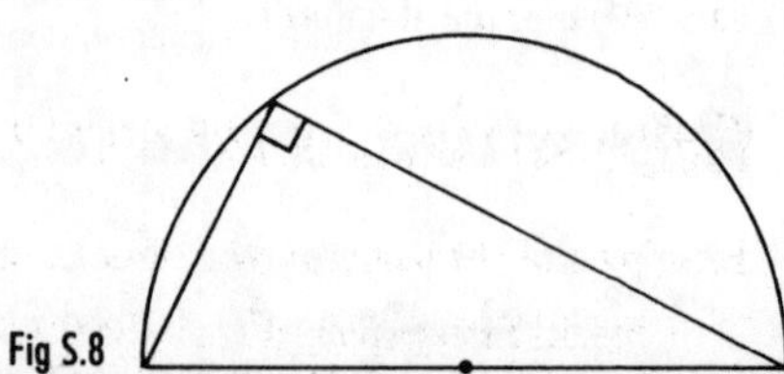

Fig S.8

SEMI-INTERQUARTILE RANGE

The semi-interquartile range is exactly half of the interquartile range.
◀ **Interquartile range** ▶

SEPTAGON

A septagon is a seven-sided plane figure, i.e. a **polygon**. It is also called a *heptagon*. A regular septagon has seven **lines of symmetry**, one through each angle bisector and **rotational symmetry** of order seven.

SEQUENCE

A sequence is a list of numbers that follow some regular, set pattern. For example:

1,4,9,16,25,36 .... n^2
2,4,6,8,10,12 .... $2n$
2,6,12,20,30 .... $(n+1) \,.\, (n+2)$

Notice how the nth term has been indicated also. Every sequence that follows a pattern will have an nth term and examination questions will often ask for this. Perhaps the best place to test this ability of spotting patterns and generalising them is within your **coursework**, and in particular during investigations.

SETS

A set is a collection of *elements* with something in common. A set is denoted in mathematics by the use of curly brackets for example {2,4,6,8} is the set of positive and even numbers less than ten.

Finite set

A finite set is like the example above, it is one with a finite number of elements.

Infinite set

An infinite set is one with an infinite number of elements, for example {positive integers} or $\{x\colon 0 \rangle x \rangle 1\}$. This last set is read as the set of numbers x such that x is bigger than 0 but less than 1. This is an *infinite set* since there are an infinite number of numbers between 0 and 1. (If you don't believe this, then just try writing all of them down, i.e. all the **vulgar fractions** as well as all the **decimals**).

SET LANGUAGE AND NOTATION

Examples of set language and notation are as follows:

$\cap$ means *intersection*; that is what is in both sets at the same time. For example $\{1,2,3\} \cap \{2,4,6,8\} = \{2\}$

$\cup$ means *union*; that is both sets combined together as one, for example $\{1,2,3\} \cup \{2,4,6,8\} = \{1,2,3,4,6,8\}$

$\mathscr{E}$ means *universal set*. This defines the limit of your situation.

$\emptyset$ means $\{\ \ \}$, in other words an *empty set*.

A′ means the *complement* of A; that is what is *not* in A. This is why we need a universal set to limit the situation. For example where $\mathscr{E} = \{1,2,3,4,5,6\}$ and $A = \{1,3,5\}$ then $A' = \{2,4,6\}$

n(A) means the *number* of A; that is how many elements are in the set A. For example in the set A above n(A)=6.

$\subset$ means a *subset of*, or is contained within. For example $\{2,3,4\} \subset \{1,2,3,4,5,6\}$

$\in$ means *is a member of*. For example $5 \in \{2,3,5,6\}$

SHEAR

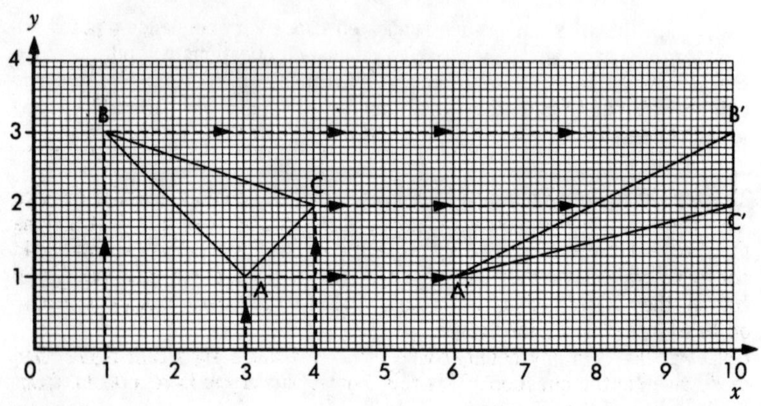

Fig S.9 Shear

A shear is a transformation that pushes a shape over in a way that the further a point is from the *invariant line*, the further it will move.

For example, suppose we wish to shear the triangle ABC in Figure S.9 with a shear factor 3, where the *x* axis is the invariant line. The diagram shows you what we have done.

Each point, and in particular each *vertex*, must move parallel to the *x* axis in a positive sense by the simple relationship:

distance from invariant line × shear factor

Hence A is to move $1 \times 3 = 3$ units
B is to move $3 \times 3 = 9$ units
C is to move $2 \times 3 = 6$ units

This moves the shape to A′B′C′ as seen in the diagram.

SIGNIFICANT FIGURES

We use significant figures whenever we estimate or make a guess at a number. A number having a specific significant figure will have that many *actual digits*, with the rest being zeros to keep place value. For example, look at the following table which illustrates various numbers of significant figures.

One significant figure	5	10	400	0.09	0.0006
Two significant figures	12	450	8300	0.37	0.029
Three significant figures	214	9560	2.65	0.0429	0.00176

We often need to **round off** to so many significant figures. When this happens we need simply to look at the first digit that 'has to go':

- if it's less than a 5 we round *down*,
- if it's a 5 or more we round *up*.

So, for example, look at the following table to see how this has been achieved for the numbers chosen:

Number	34.87	159.2	10982
One significant figure	30	200	10000
Two significant figure	35	160	11000
Three significant figures	34.9	159	11000

This topic is seldom examined on its own; it will come into a question as 'give your answer to two significant figures'. If you don't follow such instructions then you will lose marks. The biggest error on this topic is that candidates do not round off because they fail to read the *accuracy limit* on a question. Or, of course, they fail to recognise a situation where they need to *choose* to round off to so many significant figures.

Golden Rule: 'if in any doubt, round off to one more significant figure than those given in the question'. This rule only applies if you have a doubt about the intention of the question.

SIMILAR FIGURES

Two shapes are said to be (mathematically) similar if all their corresponding angles are equal and the ratios of the corresponding lengths are also equal.

For example; in Figure S.10 all the *corresponding angles*, as you can see, are equal and the ratio of each pair of *corresponding sides* are equal at 1:3.

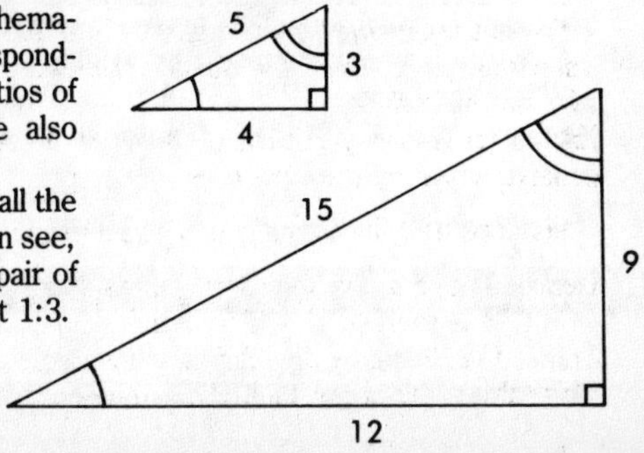

Fig S.10 Similar figures

Ratios of similar shapes

For a pair of similar shapes there is always the following relationship between the *ratios* of lengths, areas and volumes. Namely:

- where the ratio of lengths is $x:y$
- then the ratio of areas is $x^2:y^2$
- and the ratio of volumes is $x^3:y^3$

This relationship is best learnt; it nearly always crops up on examination papers somewhere.

As an example, in the Ecclesall Badminton league the winning team has a silver cup weighing 6kg and 30cm high. Each member gets a silver replica of the cup, each one 10cm high. The full size cup and the replicas are similar shapes; what is the weight of each replica?
The ratio of the *lengths* is 10:30 which simplifies down to 1:3; hence the ratio of the *volumes* (which is what we need to consider the weight) is $1\times1\times1:3\times3\times3$ which is 1:27.

Hence the weight of the replica will be 6kg divided by 27 which gives 0.2222222, which we shall round off to 0.22kg or 220g.

- **Exam Questions**
  The model railway gauge called N-gauge is a scale of 1:160. The platform of a certain railway station is 85m long. Calculate, correct to the nearest millimetre, the length of the model of the platform in N-gauge.
  (NEA; H)

- **Solutions**
  $85 \div 160 = 0.53125$cm = 531.25mm = 531mm (rounded off).

SIMPLE INTEREST

Simple Interest (SI) is the amount of money given as a result of leaving a sum of money with a bank, etc., for a particular time. It is calculated on the basis of having a ***principal amount*** P, in the bank for a number of years T, with a rate of interest R. There is then a simple formula to work out the amount of simple interest the money will earn: $\text{S.I.} = \frac{P\times R\times T}{100}$.

For example, Michael put £5 in a society that paid him 9% simple interest.
Calculate how much interest he would earn on this amount in 10 years.
Where the principal P is £5, the time T is 10 years and the rate R is 9%, then using the formula (SI = PRT/100) gives us $\text{S.I.} = \frac{5\times10\times9}{100} = £4.50$

◀ Compound interest ▶

SIMPLIFICATION

This is the process of making an algebraic expression simpler. It consists of either **factorising**, or multiplying out and collecting 'like' terms. In other words, doing anything that makes the expression simpler. For example:

- $3x-9y$ would simplify to $3(x-3y)$
- x^2+6x+5 would simplify to $(x+5)(x+1)$
  These two have been *factorised.*
- $5(x+2y)+2(4x-3y) = 5x+10y+8x-6y = 13x+4y$
  This has been *expanded* and then simplified.

- **Exam Questions**
  Simplify $3(2x-1)-2(3x-4)$.

(NEA; I)

- **Solutions**
  $6x-3-6x+8=5$

SIMULTANEOUS EQUATIONS

Simultaneous equations are a pair (at least) of equations that need solving at the same time so that the solution satisfies both of them.

The technique is simply to eliminate one variable and to solve for the single variable in the resulting simplified equation. Then to substitute that value back into one of the given equations in order to solve for the other variable. Here are two examples:

1 Solve $3x+y=11$ . . . . . . (i)
$4x-y= 3$ . . . . . . (ii)
by adding the two equations, we eliminate y
$7x = 14$
hence $x = 2$
substitute this value into equation (i)
to give $6 + y = 11$
hence $y = 5$
so the solution is $x = 2$, $y = 5$
This should be checked in equation (ii) as $8 - 5 = 3$, which is correct.

2 Solve $5x+4y = 21$ . . . . . . (i)
$3x+2y = 12$ . . . . . . (ii)
we need to multiply equation (ii) by 2
$6x+4y = 24$ . . . . . . (iii)
subtract equation (i) from equation (iii) to eliminate y
$x = 3$
substitute into equation (i)
to give $15+4y = 21$
$4y = 6$
$y = 1.5$
Check the solution in equation (ii) as $9 + 3 = 12$, which is correct.
Hence our solution is $x = 3$ and $y = 1.5$

◀ Equations ▶

- **Exam Questions**
  The first three hexagonal numbers 1, 7 and 19 are shown in the dot patterns in Figure S.11.

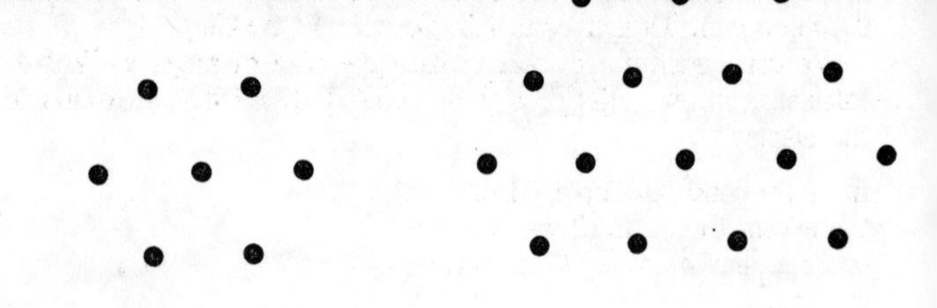

Fig S.11

a) Find the fourth hexagonal number.

b) The nth hexagonal number is given by an expression of the form an^2+bn+1, where a and b are numerical constants. Using the information above, write down **two** simultaneous equations in a and b and solve them to find a and b.

c) Verify that the 7th, 9th, 10th and 16th hexagonal numbers all contain the same three digits.

(NEA; H)

■ **Solutions**

a) 37

b) When $n=1$, $an^2+bn+1=1$, hence $a+b=0$
When $n=2$, $an^2+bn+1=7$, hence $4a+2b=6$
i.e. $a+b=0$ . . . . . . (i)
$2a+b=3$ . . . . . . (ii)
subtract (i) from (ii) to give $a=3$
substitute in (i) to give $b=-3$

c) Hence the nth hexagonal number is $3n^2-3n+1$, so
when $n=7$, $3n^2-3n+1 = 127$
when $n=9$, $3n^2-3n+1 = 217$
when $n=10$, $3n^2-3n+1 = 271$
when $n=16$, $3n^2-3n+1 = 721$
each one uses the digits 1, 2 and 7.

SINE

Sine is a trigonometrical ratio that every angle has. In a *right-angled triangle*, the sine is defined for an angle by the side opposite divided by the hypotenuse:

■ $\text{Sine} = \dfrac{\text{Side opposite}}{\text{Hypotenuse}}$

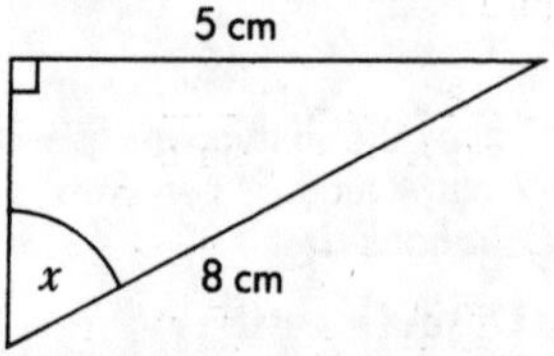

Fig S.12

For example from the lengths seen in the triangle in Figure S.12, the sine of angle x (called sin x) is usually found by dividing 5cm (side opposite) by 8cm (hypotenuse). This gives 0.625. So sin x = 0.625.

We can use this fact to calculate the size of angle x. With 0.625 in the calculator, press INV sin (or $\sin^{-1}$) and the angle should appear on the display.

In a second example, from the triangle in Figure S.13 we can state that the sine of angle 35 is y divided by 6cm.
Hence $y/6 = \sin 35$, and $y = 6 \sin 35$.

Calculate this in the calculator by putting in 35, pressing sin , then multiplying by 6 to give us 3.4414586, which we would round off to 3.4cm.

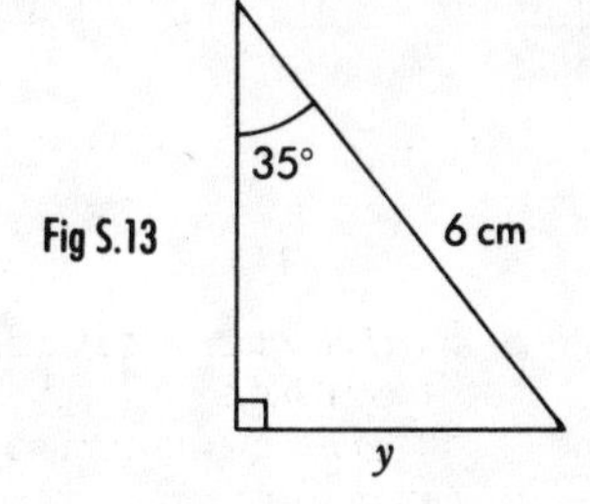

Fig S.13

SINE CURVE

The graph in Figure S.14 is of $y = \sin x$ and shows the sine curve. It is symmetrical about $x = 90°$ and $x = 270°$, hence you will see that what is drawn is only *part* of the whole curve. The sine curve continues in this form along both the positive and the negative x axis.

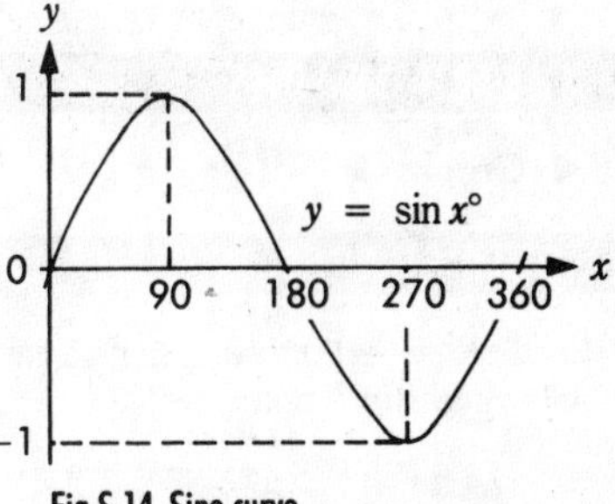

Fig S.14 Sine curve

This symmetrical pattern should help you to see that if you are given an angle, A, greater than 90°, then its sine will be the same as sin (180−A).

Your calculator will automatically give you the right sine for the angle you put into the calculator. But if, for example, you had to find what angle has a sine of 0.325, then the calculator will only tell you 19° (rounded off). There is of course another angle (180−19) which is 161°. You may realise that in fact there are lots more angles that have the sine of 0.325, which can be found by continuing the sine curve beyond 360°; but this is A level work!

SINE RULE

The sine rule works in any triangle and is illustrated on the triangle shown in Figure S.15:

$$\frac{a}{\sin A} = \frac{b}{\sin B} = \frac{c}{\sin C}$$

or

$$\frac{\sin A}{a} = \frac{\sin B}{b} = \frac{\sin C}{c}$$

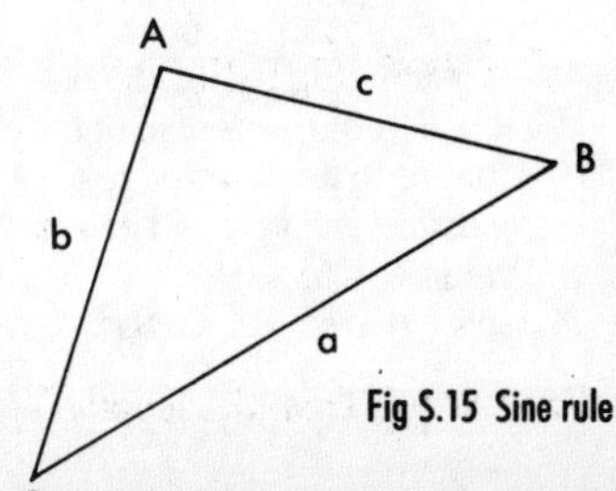

Fig S.15 Sine rule

The sine rule is used either way up, as necessary, when you are given 3 pieces of information about two angles and their opposite sides. Use the rule in such a way that the 'thing' you're looking for appears first in your formula. For example, in Figure S.16, to find the length of side c, use the rule as:

$$\frac{c}{\sin 52} = \frac{8}{\sin 69}$$

hence $c = \dfrac{8 \times \sin 52}{\sin 69} = 6.75$ cm

Fig S.16

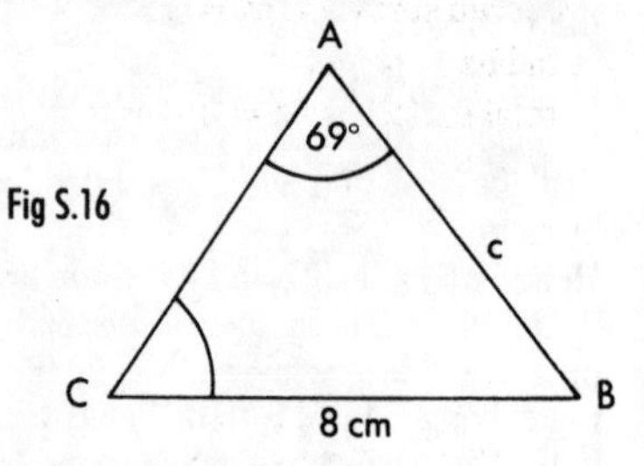

The sine rule is only used in triangles that are *not* right-angled.

SKETCH GRAPHS

◀ Graphs ▶

SOLID SHAPES

Solid shapes are three-dimensional figures, and you should be familar with the following names and facts:

Fig S.17 Cube

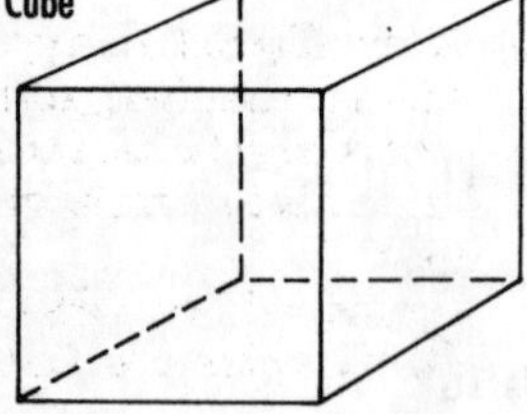

Fig S.18 Cuboid

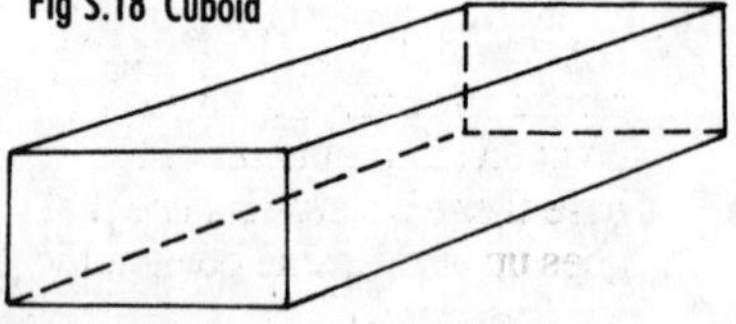

- A *cube* (Fig S.17) has all its sides the same length; each face is a square; volume=length^3
- A *cuboid* (Fig S.18) has each opposite edge the same length; each face is a rectangle; volume = length × breadth × height.
- A *sphere* (Fig S.19); the distance from the centre of the sphere to its outer edge is *constant* and is called its *radius* (r); volume = $\frac{4}{3} \times \pi \times r^3$; surface area $= 4 \times \pi \times r^2$

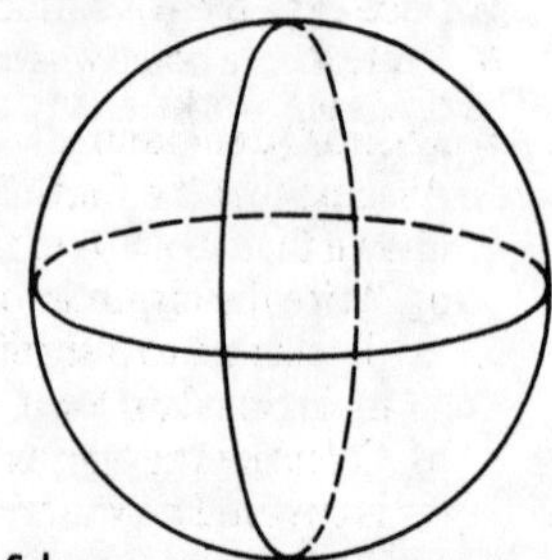

Fig S.19 Sphere

- A *cylinder* (Fig S.20) is a prism whose regular cross section is a circle;
  volume = $\pi \times \text{radius}^2 \times \text{height}$,
  curved surface area = $2 \times \pi \times \text{radius} \times \text{height}$
  total surface area = $2 \times \pi \times \text{radius} \times \text{height}$ *plus* $2 \times \pi \times \text{radius}^2$

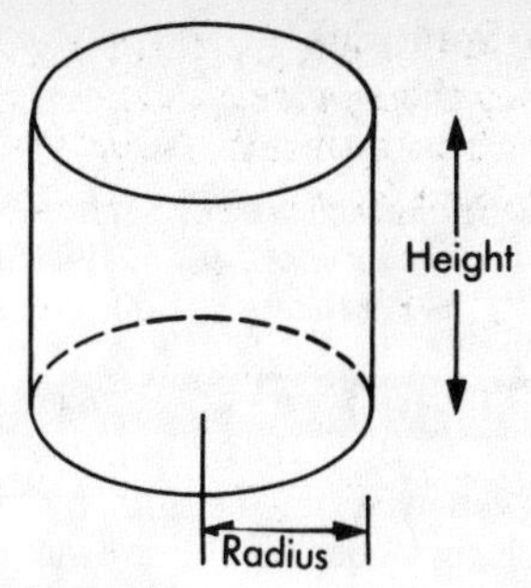

Fig S.20 Cylinder

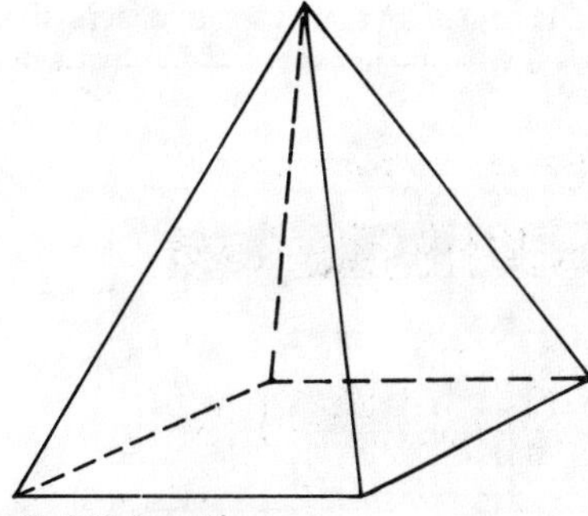
Fig S.21 Pyramid

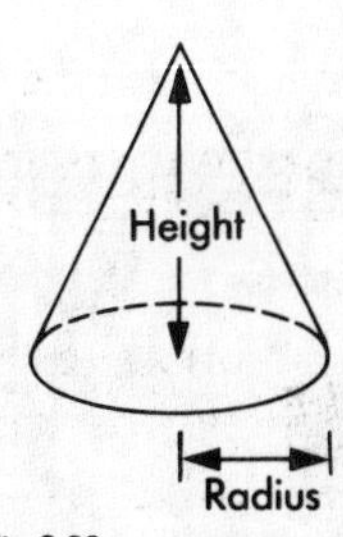

Fig S.22

- A *pyramid* (Fig S.21); the base can be any shape; from each point on the perimeter of the base there is a straight line that goes up to the same point at the top (the *vertex*);
  volume = $\frac{1}{3} \times \text{base area} \times \text{height}$.

- A *cone* (Fig S.22) is a pyramid with a circular base; where a *cone* of height h has a base radius r and a slant height of l, then:
  volume = $\frac{1}{3} \times \pi \times r^2 \times h$
  curved surface area (CSA) = $\pi r l$

Words connected with solids are *edge, face* and *vertex* (plural *vertices*):

- Edge . . . a line where two faces meet.
- Face . . . the flat surfaces of solid shapes.
- Vertex . . a point where two or more edges meet.

- **Exam Questions**

  Assume that the Earth is a sphere having diameter of 12756 km. It is known that about ⅔ of the Earth's surface is covered by water.

  a) Write the diameter of the Earth
     i) correct to 3 significant figures,
     ii) in Standard form, correct to 3 significant figures.

  b) Calculate the approximate area (in km^2) of the Earth's surface which is covered by water.

  (NEA; H)

■ Solutions

a) i) 12800 km
   ii) 1.28×10^4
b) Surface area of a sphere $= 4\pi r^2$
   $= 4 \times \pi \times 6288^2 = 496861024$
   ⅔ of this $= 331\,000\,000$ km^2

SOLUTION SETS

A solution set is the set of points that are the solution to some problem involving inequalities. For example, in Figure S.23, a solution set could be all the points in the region unshaded. It is usual to make solution sets the *unshaded* region as it is then easier to see what points there are in that set.

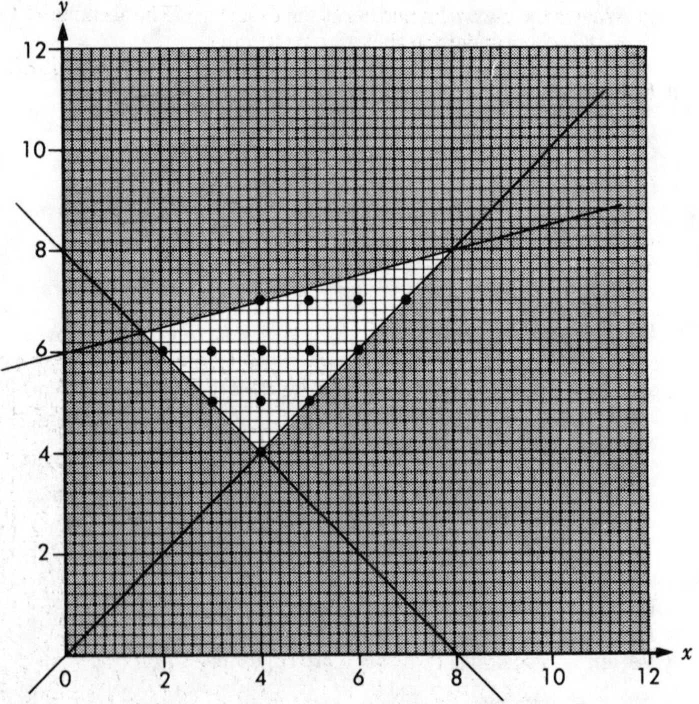

Fig S.23

■ Exam Questions

An education authority plans to build a school in a new housing estate. There are two types of teaching area, classrooms which each occupy 100m^2 and practical rooms which each occupy 130m^2.

The maximum area allocated in total for these two types of rooms is 3900m^2.

a) If x is the number of classrooms and y the number of practical rooms show that $10x + 13y \leqslant 390$.

b) A classroom accommodates 35 pupils and a practical room 20 pupils. At least 700 pupils are to be accommodated. Show that $7x+4y \geqslant 140$.

c) The number of classrooms must not exceed the number of practical rooms by more than 5.
Illustrate this condition by an inequality.

d) The number of classrooms must be at least 13.
Illustrate this condition by an inequality.

e) Using a scale of 2 cm to represent 5 units on each axis, illustrate these four inequalities by a suitable diagram on graph paper.

f) From the solution set find
 i) the minimum and maximum number of classrooms which could be built
 ii) the solution (x, y) which gives the maximum number of teaching areas.

g) What is the maximum number of pupils that could be accommodated in the school under the above four conditions?

(NISEC; H)

■ Solutions

a) $100x+130y \leqslant 3900$
divide throughout by 10 to give $10x+13y \leqslant 390$

b) $35x+20y \geqslant 700$
divide throughout by 5 to give $7x+4y \geqslant 140$

c) $x-y \leqslant 5$

d) $x \geqslant 13$

e) See Figure S.24; the unshaded region is the solution set.

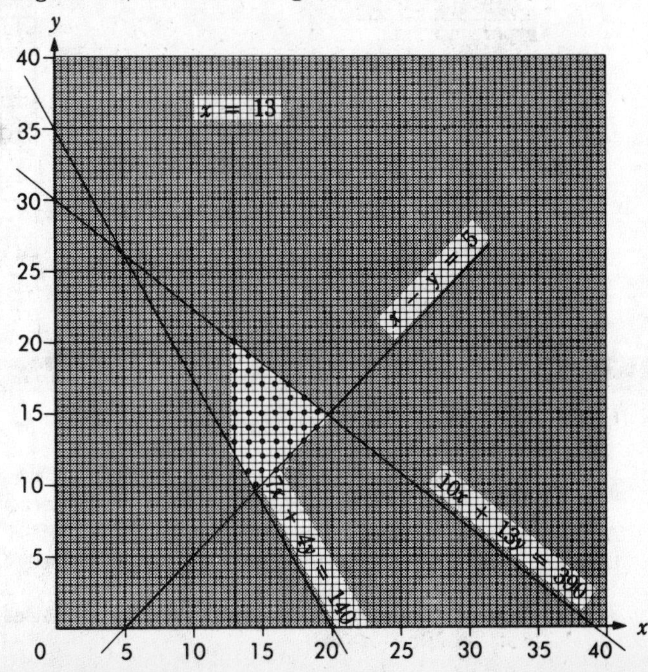

Fig S.24

f) i) minimum number = (14,11) or (15,10) = 25
maximum number = (19,15) = 34
ii) the solution (19,15)
g) maximum pupils is still at the point (19,15), which is
$(35\times19)+(20\times15)=965$

SOLVE

To solve a ***problem*** is to find the answer to it. To solve an ***equation*** is to find the number that satisfies the solution.
For example, solve $4x + 3 = 23$.
The only value of x that satisfies this equation, and therefore makes the statement correct, is $x = 5$.

■ Exam Questions

a) Basil, a window cleaner, put his 5-metre long ladder against a vertical wall to make an angle of 75° with the ground, as in Figure S.25. How far up the wall did the ladder reach?
b) If he placed the foot of the ladder 90cm away from the wall, how far up the wall would his ladder reach?

(NEA; I)

■ Solutions

a) Recognise the trigonometry sine situation, to give:

$$\frac{\text{opposite}}{\text{hypotenuse}}$$

$$\frac{\text{opposite (height)}}{5} = \sin 75°$$

hence height = $5 \sin 75°$ = 4.83 or 4.8m.

b) Recognise the Pythagoras situation, to give
$\text{height}^2 = 5^2 - (0.9)^2$
$= 24.19$
height $= \sqrt{(24.19)} = 4.92$
or 4.9m.

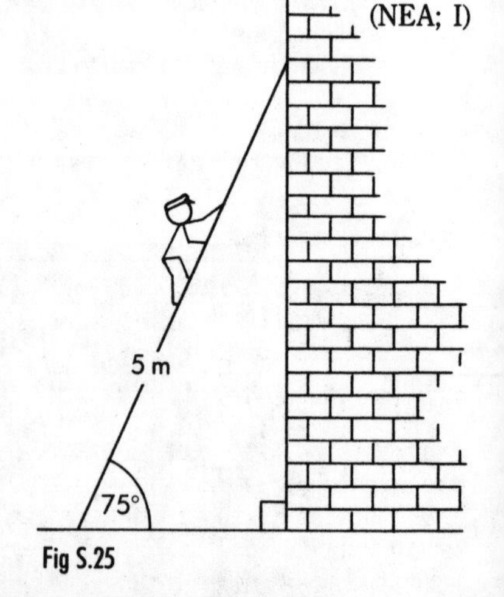

Fig S.25

SPEED

Speed is the rate of change of distance. It can be found by the ***gradient*** of a distance/time graph.

Average speed

Average speed is found by the total distance travelled divided by the total travelling time. The ***units*** of speed can vary with the data used to calculate it.
For example James cycled 28 mile in 2 hours, or $^{28}/_{2}$ = 14 miles per hour.
John ran 10000 metres in 14 minutes, or $^{10000}/_{14}$ = 714 metres per minute.

SPHERE

A sphere is a solid shape. Every point on its outer surface is the same distance from the centre of the sphere. A sphere is a ball-shape, and the shape of the world is used as a sphere for the sake of calculations.

- The volume of a sphere is equal to $\frac{4}{3} \times \pi \times \text{radius}^3$
- The surface area of a sphere is equal to $4 \times \pi \times \text{radius}^2$

◀ Solid shapes, Sphere ▶

SQUARE

A square has all its four sides equal and all its angles are right angles. A square has **rotational symmetry** of order 4, because if you turn it round its centre, there are four different positions it can take that all look the same, as shown in Figure S.26.

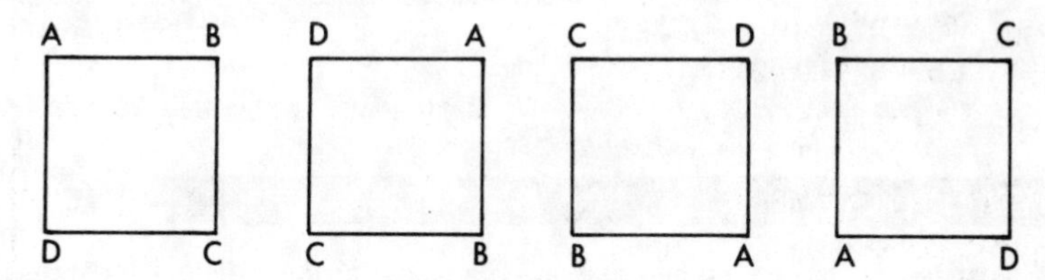

Fig S.26

Square numbers

A square *number* is one that has a **factor** which can multiply by itself to give this square number. For example, 16 has the factor 4 and we know that $4 \times 4 = 16$. The first few square numbers are 1,4,9,16,25,36,49,64 and 81.

Square roots

The square *root* of a number is that number that multiplies by itself to give you that number For example, the square root of 9 is 3, since $3 \times 3 = 9$.

Of course if a number has an **integer** square root, then this number is a square number.

You find square roots on your calculator by pressing the button with $\boxed{\sqrt{\ }}$ on it. Every number has two square roots, one positive the other negative. For example, the square roots of 25 are 5 and -5.

STANDARD FORM

Standard form is a convenient way of writing very large or very small numbers. It is always expressed in the terms of:

$a \times 10^N$

Where a is a number between 1 and 10, and N is an integer.

For example:

400	would be written as	4.0×10^2
35687	would be written as	3.5687×10^4
132.98	would be written as	1.3298×10^2

Note how the index on the 10 tells you how many places to move the decimal point. If the number is *less* than 0 to start with then the index will be *negative*. For example:

0.000000037	would be written as	3.7×10^{-8}
0.0020005	would be written as	2.0005×10^{-3}

■ **Exam Questions**

A pack of 52 playing cards is 1.5 cm thick. Write down, in Standard form, the thickness, in cm, of one playing card.

(NEA; I)

■ **Solutions**

$1.5 \div 52 = 0.0288$

$= 2.88 \times 10^{-2}$

STATISTICS

Statistics are pieces of information. Statistical *methods* help us express information in a way that makes the information clear. This may involve using charts and diagrams, and making simple calculations.

Questions are always asked on statistics in your examinations and, if you read the question carefully, then they should be relatively easy questions to answer.

◀ Bar charts, Pictograms, Pie charts, Histograms, Averages, Frequency ▶

STRETCH

A stretch is a geometrical transformation, which enlarges a shape in different directions with different 'stretch factors'.

One way stretch

This is where we stretch in *one direction* only. For example in Figure S.27, the triangle T has undergone a one-way stretch of scale factor 3 with the *x-axis invariant* to give the shape TT.

Notice each point in triange T has been moved perpendicular to the invariant line and to a point a distance away from the invariant line found by 'distance from invariant line multiplied by the scale factor.'

Two way stretch

This is where we stretch in *two directions* at once. For example in Figure S.28, the triangle T has undergone a two way stretch of scale factor 3, with the *x-axis invariant* and scale factor 2, with the *y-axis invariant*. This gives the shape TTT.

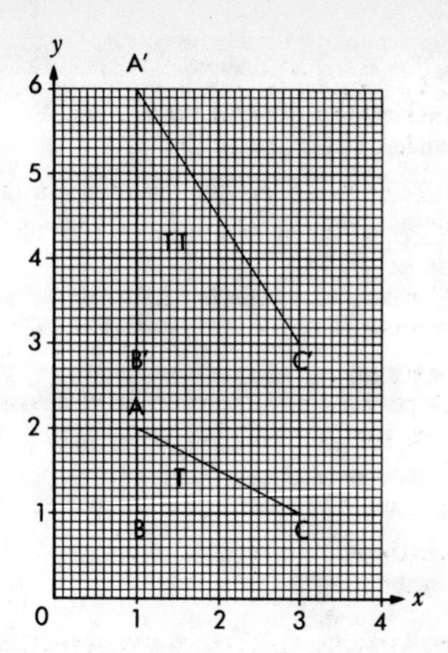

Fig S.27 One way stretch

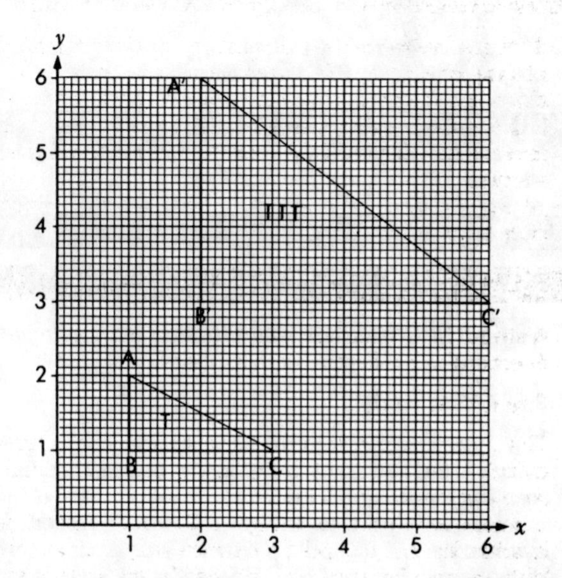

Fig S.28 Two way stretch

Notice that each point (x, y) on the triangle T has moved to a point such that:

- new x ordinate is x multiplied by 2
- new y ordinate is y multiplied by 3

Where 2 and 3 are the relevant *scale factors*.

SUBSET

A subset is a set contained within another set. For example a subset of {1,2,3,4,5,6} could be {1,2,3} or even {5}.

For example, the subsets of {a,b,c} are { }, {a}, {b}, {c}, {a,b}, {a,c}, {b,c} and {a,b,c}.

Note that the complete set itself, as well as the empty set, are also considered as subsets.

A set with N elements in will have 2^N different subsets.

The symbol for subset is C. For example {7,8} C {6,7,8,9}.

SUBSTITUTION

Substitution is where you put a particular value in place of a variable in an expression or a formula. For example, if I substitute $x=3$ into the equation $y=5x+2$, then the value of y will become $y=5\times3+2=17$.

- **Exam Questions**

  Calculate the exact value of x^2-4x-1

  a) when $x=-1$,

  b) when $x=2.56$.

- **Solutions**

  a) $1--4-1 = 1+4-1=4$

  b) $5.1076-10.24-1=-4.6864$

SUBTRACTING/SUBTRACTION

◀ Directed numbers, Matrices ▶

SYMMETRY

Symmetry is found in two and three dimensional shapes.

TWO-DIMENSIONAL SYMMETRY

There are two particular types of symmetry:

Line

If you can fold a shape over so that one half fits exactly on top of the other half, then the line over which you have folded is called a **line of symmetry**. The examples in Figure S.29 illustrate this. The dotted lines are lines of symmetry.

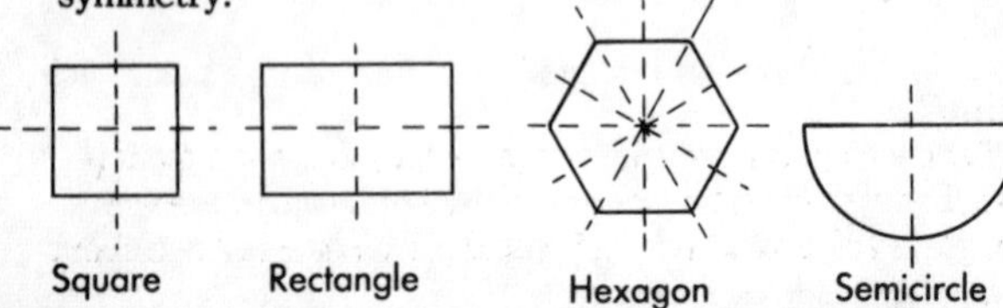

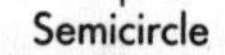

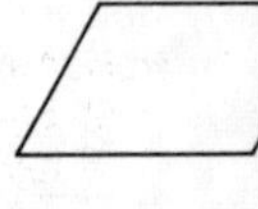

Fig S.29

Rotational

This is sometimes called *point symmetry*. A shape has rotational symmetry according to how many different positions it can be turned round to, so as to look exactly the same.

For example, a square has rotational symmetry of order 4, since if you turn it round its centre there are four different positions that it can take that all look the same, as shown in Figure S.30.

Look at the shapes in Figure S.31; their orders of rotational symmetry have been given. Any shape that has what we call 'no symmetry', such as the letter Q, has rotational symmetry of *order 1*, since there is only *one* position in which it looks the same.

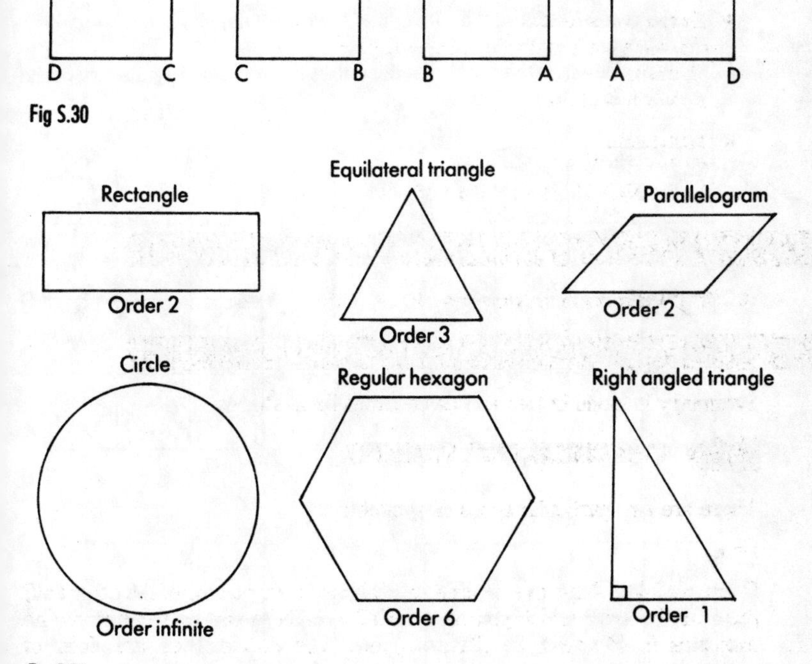

Fig S.30

Fig S.31

■ **Exam Questions**

a) Figure S.32 shows four square tiles with a pattern drawn on the top left hand tile. Draw the patterns needed on the other three tiles so that the completed picture is symmetrical about the line *AB* and about the line *CD*.

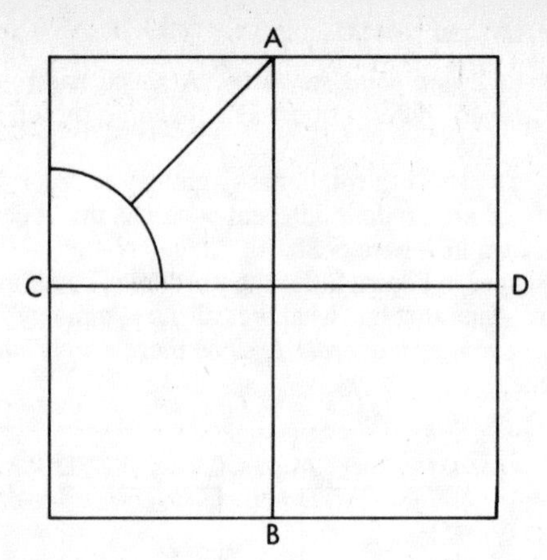

Fig S.32

AB and CD intersect at the point O, which is not labelled in the diagram.
Write down a statement that describes the rotational symmetry of the finished pattern about O.

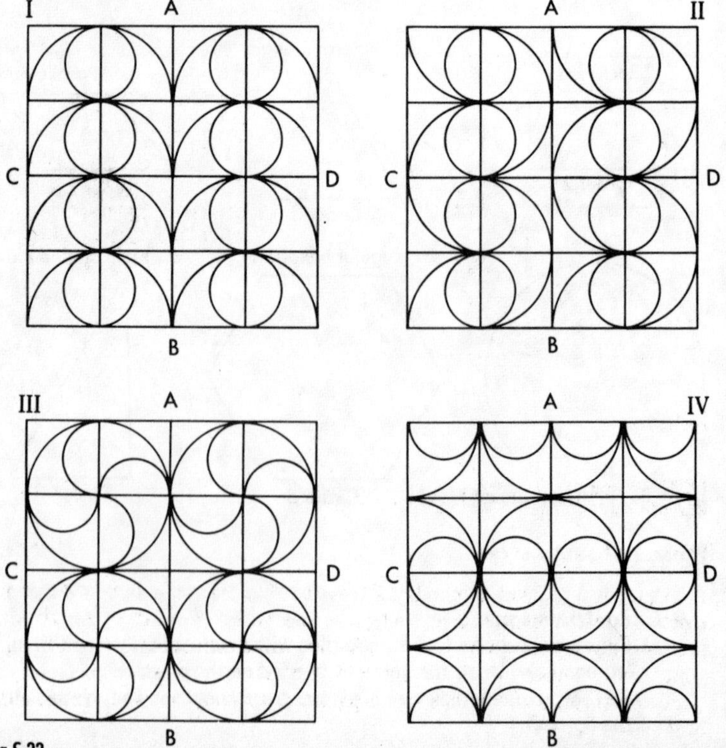

Fig S.33

b) Look at the four patterns in Figure S.33. In each pattern, O is the point of intersection of AB and CD.
Pattern IV has a tick in the box for "Rotational symmetry of 180° about O, because if the pattern is rotated about O through 180° then the pattern would look unchanged.

Pattern	*Symmetrical about AB*	*Symmetrical about CD*	*Rotational Symmetry of 180° about O*
I			
II			
III			
IV			✓

For each pattern in turn, place a tick (✓) in the box if the pattern has the property. A pattern may have more than one of the symmetries given in the table.
(WJEC; I)

■ **Solutions**

a) See Figure S.34; rotational symmetry of order 2.

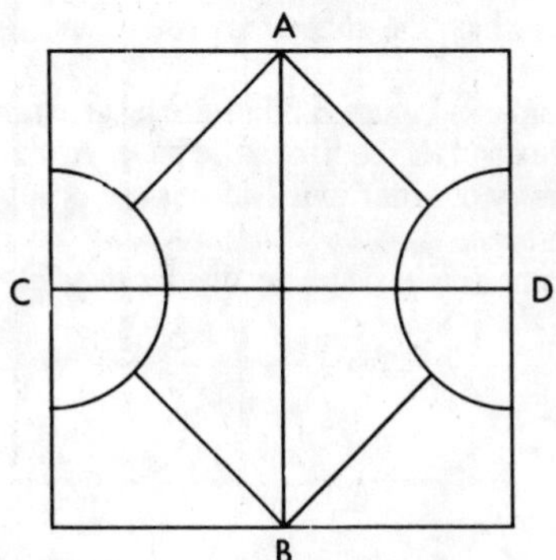

Fig S.34

b)

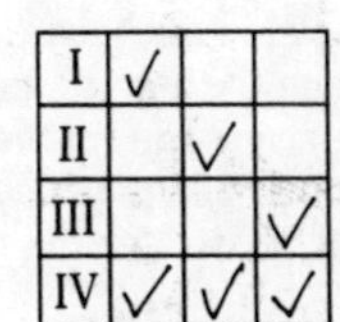

I	✓		
II		✓	
III			✓
IV	✓	✓	✓

THREE DIMENSIONAL SYMMETRY

Planes of symmetry

A shape has a plane of symmetry if you can slice the shape into two matching pieces, one the exact mirror image of the other. To find these planes of symmetry you need to be able to visualise the shape being cut and to see in your mind whether the pieces are matching mirror images or not.

For example, a cuboid has three planes of symmetry, as shown in Figure S.35.

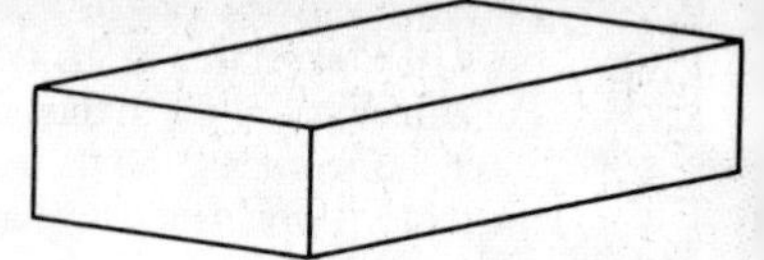

Find how many planes of symmetry a cuboid has

Consider the cuboid shown above. We can cut it into two exact halves in the following three ways. Hence the shape has three planes of symmetry.

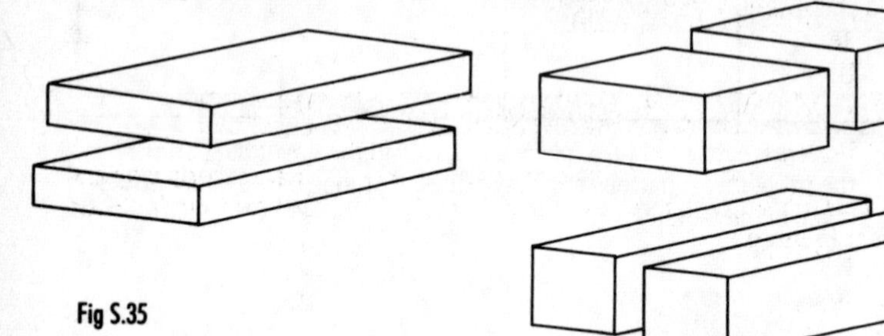

Fig S.35

Axes of symmetry

An **axis of symmetry** is a line through which the shape may rotate and yet still occupy the same space.

For example, in the square based pyramid in Figure S.36, there is an *axis of symmetry* along the line through the vertex and the centre of the base. Around this axis the shape has **rotational symmetry** of order four, since it can occupy four different positions within the same space.

Figure S.37 illustrates the three different *axes of symmetry* that a *cuboid* has.

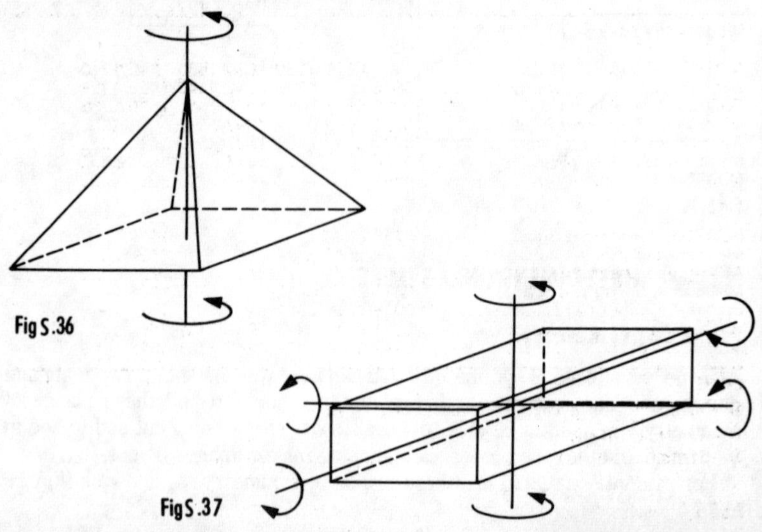

Fig S.36

Fig S.37

TABLES

Tables are 'containers' of information which are set out in an orderly way and are largely numeric. There are a number of tables that you ought to be confident in using.

TYPES OF TABLES

Timetables

Figure T.1 shows an example of a timetable. You should be able to work out times of buses from one place to another as well as be able to plan out a journey using timetables like these.

SHEFFIELD CITY SERVICE — **ONE MAN Service**

CITY · GREYSTONES · NORTON — **40**

Monday to Friday only

	⊗			⊗	□		
ARUNDEL GATE	....	0820	1230	1530	1530	1610	1710
Hunters Bar	0805	0835	1245	1545	1545	1625	1725
Ecclesall, Knowle Lane	0816	0846	♥	1556	♥	1636	1736
Millhouses, Springfield Road	0822	0852	....	1602	....	....	....
Abbey Lane, Bocking Lane	0827	....	....	1607	....	....	....
Meadowhead, Norton Hotel	0834	....	....	1614	....	....	....
NORTON, Cloonmore Drive	0837	....	....	1617	....	....	....

		⊗	□			⊗	□
NORTON, Cloonmore Drive	0755	0840	....	....	....	1521	....
Meadowhead, Norton Hotel	0758	0843	....	....	....	1524	....
Abbey Lane, Bocking Lane	0805	0850	....	....	....	1531	....
Millhouses, Springfield Road	0810	0855	0855	....	....	1536	....
Ecclesall, Knowle Lane	0816	0901	0901	●	●	1542	●
Hunters Bar	0826	0911	0911	0941	1251	1557	1557
ARUNDEL GATE	0842	0927	0927	0957	1307	1613	1613

CODE

⊗ – Runs Schooldays only. □ – Not Schooldays.

♥ – Runs to Rustlings Road Junction.

● – Starts from Rustlings Road Junction 1 minute before time shown for Hunters Bar.

Fig T.1 Timetable

Tidetables

The tidetable in Figure T.2 shows the approximate times of the high tides for the first ten days in February 1989. You will read that the high tide on February 8th was at 0829, which is twenty nine minutes past eight in the morning, and 2046, which is fourteen minutes to nine in the evening.

	February 1989						
	Morning			Afternoon			
Date	*Time*	*Height Mtrs*	*Height Ft ins*	*Time*	*Height Mtrs*	*Height Ft ins*	*Moon*
1	0449	3.20	10.5	1711	3.20	10.5	
2	0526	3.60	11.8	1746	3.51	11.5	F.M.
3	0600	3.75	12.3	1818	3.60	11.8	
4	0631	3.87	12.7	1848	3.60	11.8	
5	0702	3.87	12.7	1917	3.60	11.8	
6	0730	3.87	12.7	1947	3.60	11.8	
S 7	0759	3.75	12.3	2015	3.38	11.1	
8	0829	3.51	11.5	2046	3.08	10.1	
9	0900	3.20	10.5	2119	2.87	9.4	
10	0936	2.87	9.4	2200	2.47	8.1	Last Qtr

Fig T.2 Tidetable

Insurance tables

You can see from the insurance table in Figure T.3 that costs vary for different ages and for the amount of insurance required.

	Monthly insurance premiums			
Age (next)	£1000	£5000	£10 000	£50 000
20	1.65	7.80	14.50	55.10
25	1.90	8.15	16.20	63.80
30	2.40	9.00	17.65	68.10
35	3.45	10.05	19.95	77.90
40	6.55	18.95	37.40	165.50
45	15.80			

For female subtract 5 years from current age

Fig T.3 Insurance table

Cost table

The table in Figure T.4 shows you the different charges for first class and second class post at a particular date.

POSTAL INFORMATION INLAND

Weight not exceeding	*First Class*	*Second Class*	*Weight not exceeding*	*First Class*	*Second Class*
60	p	p	350 (12.3 oz)	61	46
100 (2.1 oz)	19	14	400 (14.1 oz)	69	52
100 (3.5 oz)	24	18	450 (15.9 oz)	78	59
150 (5.3 oz)	31	22	500 (1.1 lb)	87	66
200 (7.1 oz)	38	28	750 (1.7 lb)	128	98
250 (8.8 oz)	45	34	1000 (2.2 lb)	170	max
300 (10.6 oz)	53	40	Each extra 250g (18.8 oz)	42	—

Registration. Minimum fee, £1; compensation up to £500.

Recorded delivery. Letters and Packets: fee, 20p in addition to postage. Compensation up to £18.

Fig T.4 Cost table

■ Exam Questions

1 Figure T.5 gives repayment details for various amounts of a loan. Some of the figures in the table have been obscured by a coffee stain.

REPAYMENT PERIOD	48 MONTHS	60 MONTHS	72 MONTHS
LOAN AMOUNT	MONTHLY REPAYMENTS (total amount repayable in brackets)		
£3000	£87.19 (£4185.12)	£75.20 [illegible]	[illegible]
£4000	[illegible] (£5580.00)	[illegible]	[illegible]
£5000	£145.31 (£6974.88)	£125.20 (£7520.40)	[illegible]
£7500	£217.97 (£10 462.56)	£188.01 (£11 280.60)	[illegible] (£12 185.22)

Fig T.5

Use the table to answer the questions below.

a) If you take a loan of £3000 and pay it back over 60 months, calculate the total amount you pay back.
b) How much would you pay back per month for a loan of £4000 over 48 months?
c) Work out the monthly repayment for a loan of £3800 over 4 years.

(NEA; H)

2 Figure T.6 shows the times of the 327 bus on Sundays.

a) What is the latest time you can catch a bus from Luton to Rickmansworth?

Sundays and Public Holidays (Bus 327)

Luton *Bus Station* ⇌	..	..	...	0850		50		1750	...	1850	....	....	....	....
Lantern Fields *Slip End Turn*	....	..	....	0859		59		1759	....	1859	..	..	..	..
Kinsbourne Green *Harrow*	..	..	....	0904		04		1804	....	1904	..	....	..	..
Harpenden *George* ⇌	..	..	..	0910	Then	10		1810	....	1910	....	....	....	....
Sandridgebury Lane	....	..	....	0920	at	20		1820	....	1920	....	....	....	....
St. Albans *Bus Garage* (A)	..	..	....	0923	these	23		1823	1917	1923	2023	2123	2217	2317
St. Albans *St. Peters Street*	....	....	....	0824	minutes	24	UNTIL	1824	1918	1924	2024	2124	2218	2318
Chiswell Green *Three Hammers*	....	....	....	0933	past	33		1833	1927	1933	2033	2133	2227	2327
Garston *Watford Bus Garage*	0720	0758	0841	0841	each	41		1841	1936	1941	2041	2141	2235	2335
Watford Junction ⇌	0731	0809	0852	0952	hour	52		1852	....	1952	2052	2152	2246	....
Watford *Town Centre*	0734	0812	0855	0955		55		1855	....	1955	2056	2155	2249	....
Croxley Green Station ⇌	..	0819	0902	1002		02		1902	....	2002	2102	2202	2256	....
Croxley Green *Manor Way*	....	0823	0906	1006		06		1906	....	2006	2106	2206	2300	....
Rickmansworth *Station* ⊖ ⇌	....	....	0915	1015		15		1915	....	2015	2115	2215	....	....

Fig T.6

b) How many minutes does the bus take to travel from St. Albans Bus Garage to Watford Junction?

c) How many journeys are made by a 327 bus on Sundays from Luton to Rickmansworth?

(SEG; B)

■ Solutions

1 a) £75.20×60 = £4512

b) £5580÷48 = £116.25

c) £1000 will cost £87.19÷3
hence £3800 will be £87.19÷3×3.8 = £110.44

2 a) 1850 or ten to seven.

b) From 23 to 52 or from 17 to 46 minutes past the hour. This gives a time of 29 minutes.

c) There are 11

TALLY CHART

A tally chart is a chart that helps us to keep count of different events during some experiment or trial.

For example, if you were doing a count of the favourite flavour crisp at your school or college you could use a chart like that in Figure T.7. Each time you asked someone their favourite flavour, you put a mark in the correct row. When you wish to put down the fifth mark you put a diagonal line across the others, like a bar on a gate as shown. This way you can more easily count them all up at the end. If someone wants a flavour that you haven't got, you put that mark in the row headed 'others'.

From the tally chart you would then construct a *frequency* table of some sort.

Flavour	*Tally*	*Total*
Plain	𝍸 𝍸 𝍸 II	17
Salt 'n vinegar	𝍸 𝍸 𝍸 𝍸 𝍸 𝍸 IIII	34
Cheese & onion	𝍸 𝍸 I	11
Others	𝍸 𝍸 III	13

Fig T.7 Tally

TANGENT

There are two uses of the word tangent in mathematics.

TANGENT IN TRIGONOMETRY

Tan, which is short for tangent, is a *trigonometrical ratio* that every angle has. The tan ratio is defined for an angle in a right-angled triangle, by the side opposite divided by the side adjacent

- $\text{Tan} = \dfrac{\text{Side Opposite}}{\text{Side Adjacent}}$

For example, from the lengths seen in the triangle in Figure T.8, the tan of angle x (tan x) is found by dividing 8cm (side opposite) by 5cm (side adjacent). This gives 1.6, so tan x = 1.6.

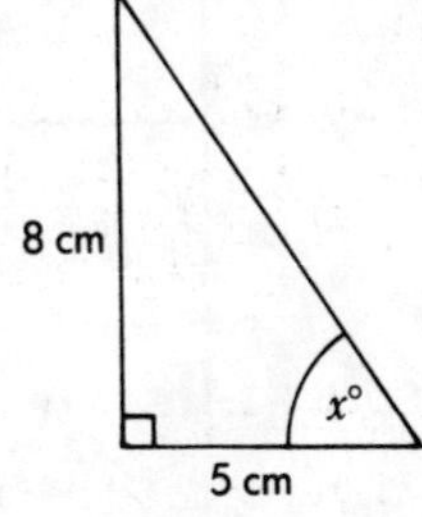

Fig T.8

We can use this fact to calculate the size of angle x. With 1.6 in the calculator, press [INV] [tan] (or [$\tan^{-1}$]) and the angle should appear on the display as 58° (rounded off).

In the triangle in Figure T.9 we can state that the tan of angle 29° is y divided by 6cm.
Hence $y/6 = \tan 29°$, so $y = 6 \times \tan 29°$.

Work this out in the calculator by putting in 29, pressing [tan] and then multiplying by 6 to get 3.325843, which would round off to 3.3cm.

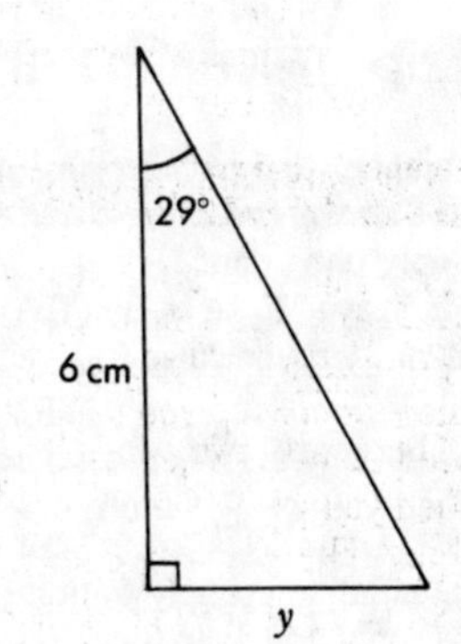

Fig T.9

Tan curve

The graph in Figure T.10 is of $y = \tan x$ and shows the tan curve. If you are given an angle, A, greater than 90°, then its tan will be the same as tan (A − 180°).

Your calculator will automatically give you the right tan for an angle you put into the calculator. But if, for example, you were to find what angle had a tan of 0.8, then the calculator would tell you 39° (rounded off). There is however another angle, 180°+39° which is 219°.

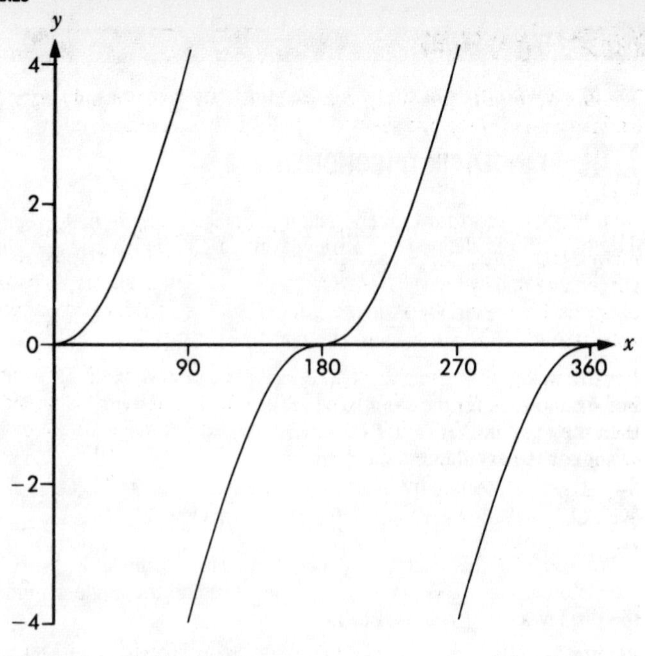

Fig T.10

TANGENT TO A CIRCLE OR CURVE

A tangent to a circle, or a curve, is a line that will touch the circle or curve at only one point.

If drawn to a circle, the tangent will be perpendicular to a radius (Fig T.11).

There are two ways to draw a tangent to a circle at a particular point. One way is to put your ruler on that point and simply to draw the line that only touches the circle there. This is in fact the only way when it comes to drawing a tangent on a *curve* (as when finding the **gradient** of a curve). The other way is to **construct** a right angle at that point on the radius and hence draw in the tangent.

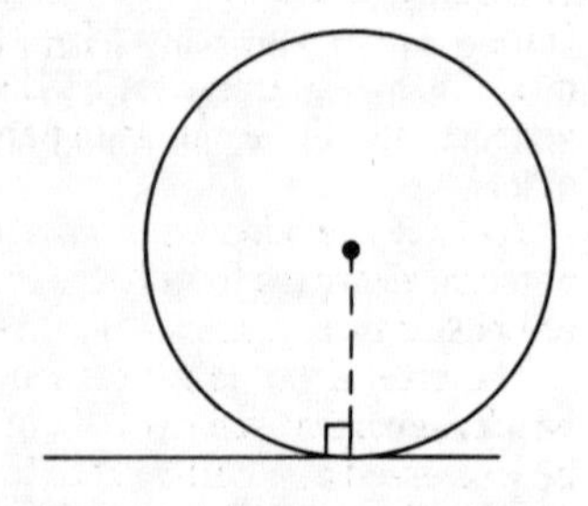

Fig T.11 Tangent

TAXES

Tax is the amount of money that a government tells its people to pay in order to raise sufficient funds for that government to run the country.

TAX SYSTEMS

The tax system is generally complicated, but you are only expected to be familiar with two major types of tax, VAT and Income Tax.

VAT

VAT, or Value Added Tax, is the tax put onto the price of goods sold in shops, restaurants, etc., then paid to the government. The tax is usually a percentage and can vary from year to year. The tax also depends on the type of goods being sold, with some goods (such as children's clothes) having a zero rate of VAT, and others the usual 15% rate.

For example, a fish and chip shop tells its customers that if they eat their food in the shop then they have to pay VAT at 15%; if they eat it outside the shop then they pay no VAT. If it costs 98p for fish and chips to eat outside, what will it cost to eat them inside?

The inside price will be 115% of 98p which is $98 \times {}^{115}/_{100}$ which is 112.7; this will round off to £1.12 (note the practice in tax to **truncate** when rounding the tax to be added or paid).

Income tax

Income tax is the type of tax that almost everyone who receives money for working or from investments has to pay to the government. Here again, the amount can change every time the government decides to change it (usually at Budget time). To calculate how much tax you should pay you first need to know the *rate of tax* (a percentage) and your *personal allowances*.

Personal allowances are the amounts of money you may earn before you start to pay tax. It will be different for married men, single men and for women in different situations, and can be increased for quite a variety of reasons. Once your personal allowance has been calculated by the tax man, he will divide it by ten, truncate the decimal fraction, and give you the number as your *tax code*. For example, a personal allowance of £2400 will have a tax code of 240.

You only pay tax on your *taxable income*, which is found by subtracting your personal allowance from your total annual income. If your personal allowances are higher than your income, then you pay no income tax.

The rate of tax is expressed either as a percentage, for example 25%, which means that you pay 25% of your taxable income as income tax, or it may be expressed at a certain rate in the £. For example, if it were 24p in the £, you would pay 24p for every £1 of your taxable income (which is equivalent to 24%).

For example, if the rate of tax is 25%, find the income tax paid by Mr Kaye who earns £11,500 per annum and has a tax code of 346.
Mr Kaye's personal allowance is $346 \times 10 = £3,460$,
hence his taxable income $= £11,500 - £3,460 = £8,040$
So the tax paid is ${}^{25}/_{100} \times £8,040 = £2,010$.

TERMINATING

◀ Decimal ▶

TESSELLATIONS

A *regular* tessellation is a regular pattern with *one shape* that could cover a large area without leaving any gaps (except at the very edge). Figure T.12 shows some examples of regular tessellations.

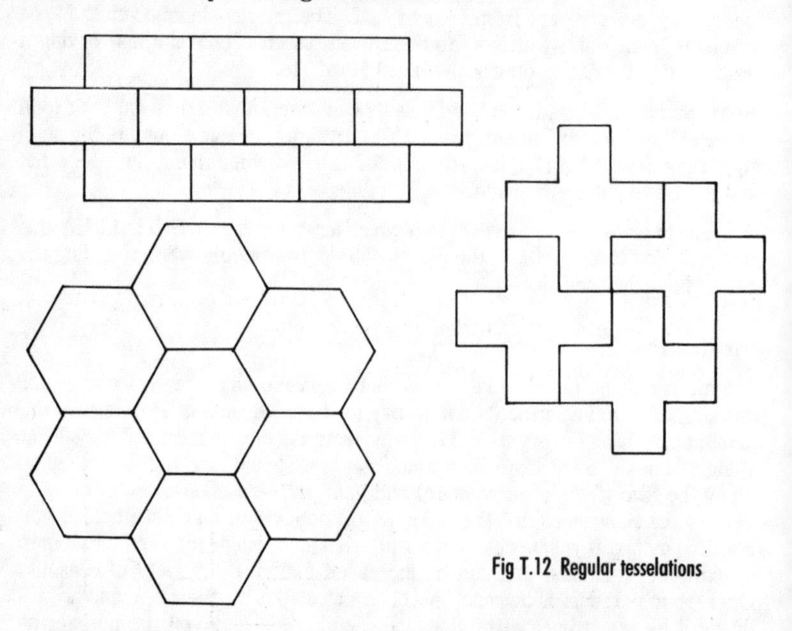

Fig T.12 Regular tesselations

Each tessellation is made from one plane shape and could continue its pattern to fill a large area without leaving any space in between. It is true to say that *every triangle* and *every quadrilateral* will tessellate.

Semi-regular tessellation

A semi-regular tessellation is one which uses *more than one shape* to create the regular pattern, again leaving no gaps, as in Figure T.13.

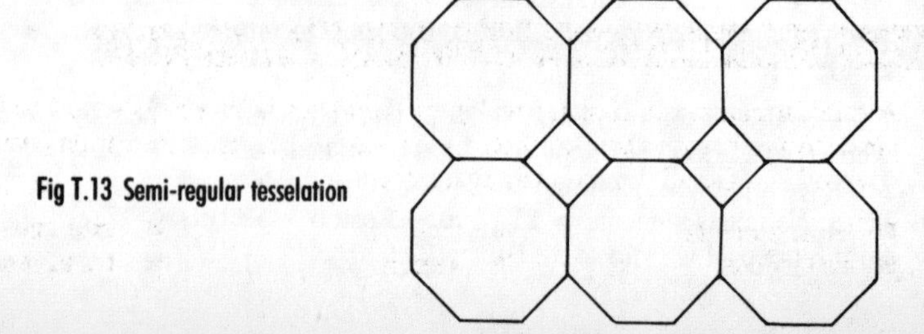

Fig T.13 Semi-regular tesselation

■ **Exam Questions**

Draw a tessellation of the given shape on the grid in Figure T.14. The shape should be repeated at least eight times.

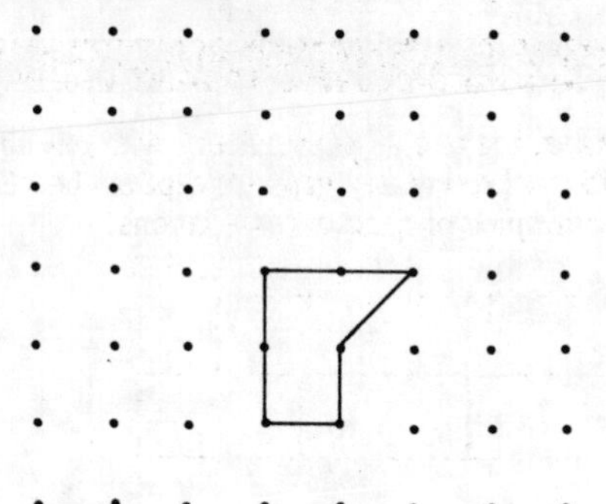

Fig T.14

■ **Solution**

See Figure T.15.

You should have shown at least 8 shapes fitting to a regular pattern as this. (There are only two possible patterns).

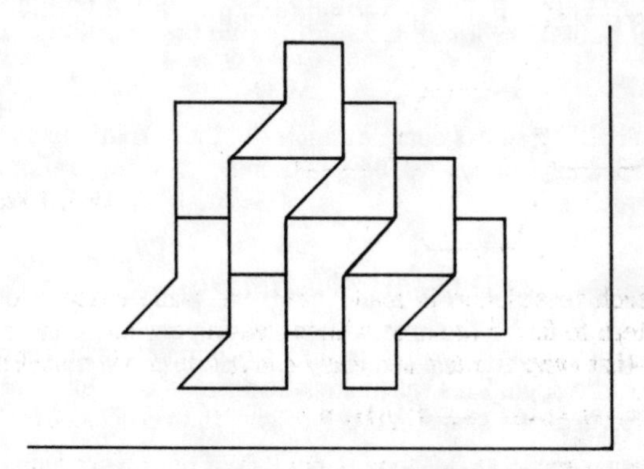

Fig T.15

TRANSFORMATIONS

A transformation is a change; within mathematics it invariably refers to how *plane shapes* change their position and/or shape. The transformations usually covered in a GCSE syllabus will result from:

- enlargements
- reflections
- rotations
- shears
- stretches
- translations

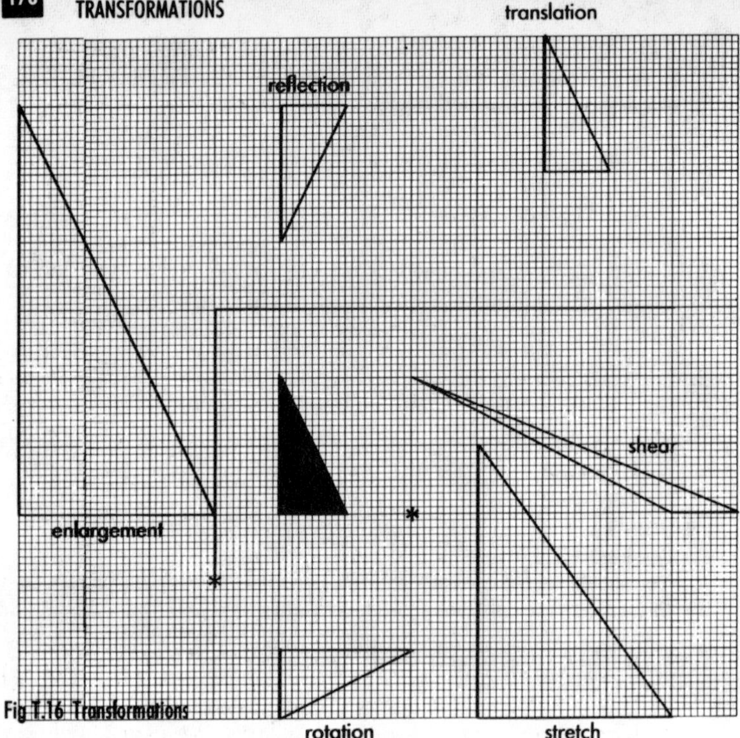

Fig T.16 Transformations

Figure T.16 shows some examples of these tranformations. However, more detailed explanation will be found under their own reference headings within this guide.

TRANSFORMATION MATRIX

A transformation matrix is a 2 by 2 matrix that represents a particular transformation. This will multiply to a matrix containing the **position vectors** of the *vertices* of a shape in order to give the vertices of the *transformed* shape.

Identity transformation

The identity transformation is the transformation that leaves a shape where it originally was; for example a **rotation** of 0° or 360°, or even an **enlargement** of scale factor 1.

Inverse transformation

The inverse transformation (T′) of a transformation T, is the transformation that will move a shape back to its original position after first being moved by T. For example, the inverse of any **reflection** is the same reflection (a **self inverse**), and the inverse of any **rotation** of angle A will be a rotation of −A around the same point.

Fig T.17

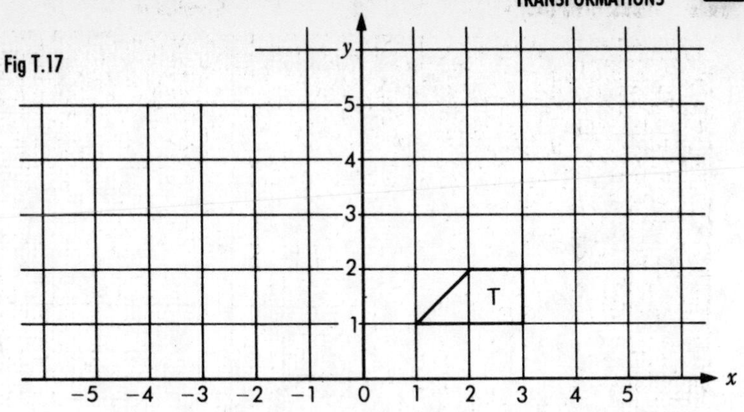

■ Exam Questions

1 Trapezium T in Figure T.17 is reflected in the line $y = 3$ to give trapezium T_1.
Trapezium T_1 is then rotated 90° anticlockwise about (0,0) to give trapezium T_2.
Trapezium T_2 is then translated $\begin{pmatrix}6\\0\end{pmatrix}$ to give trapezium T_3.

a) Show the positions of T_1, T_2 and T_3 on the diagram.
b) Describe the single transformation that would take T to T_3.

(NEA: I)

2 Look at Figure T.18. Describe fully two different single transformations which will map the right hand A on to the left hand A.

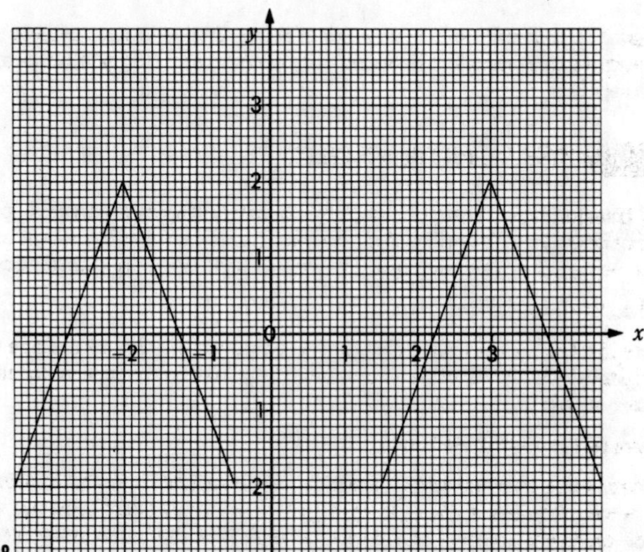

Fig T.18

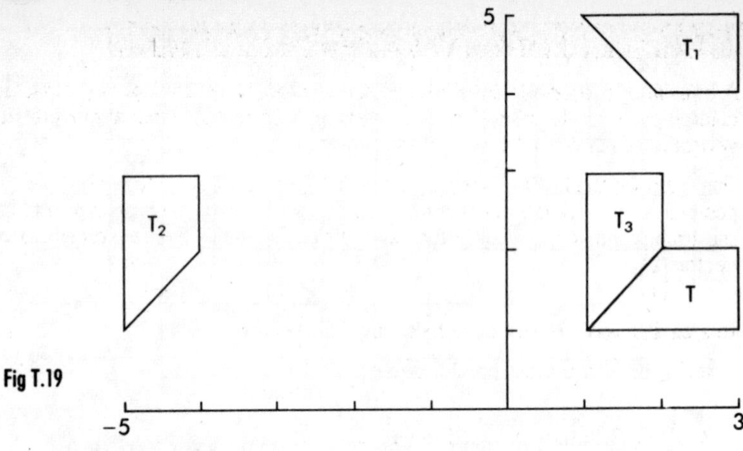

Fig T.19

■ Solutions

1 a) See Figure T.19
1 b) A reflection in the line $y=x$

2 i) A reflection in the line $x=\frac{1}{2}$
2 ii) A translation with the vector $\begin{pmatrix}-5\\0\end{pmatrix}$

(The most common error here would be to misread the question and give the translation as left to right instead of right to left).

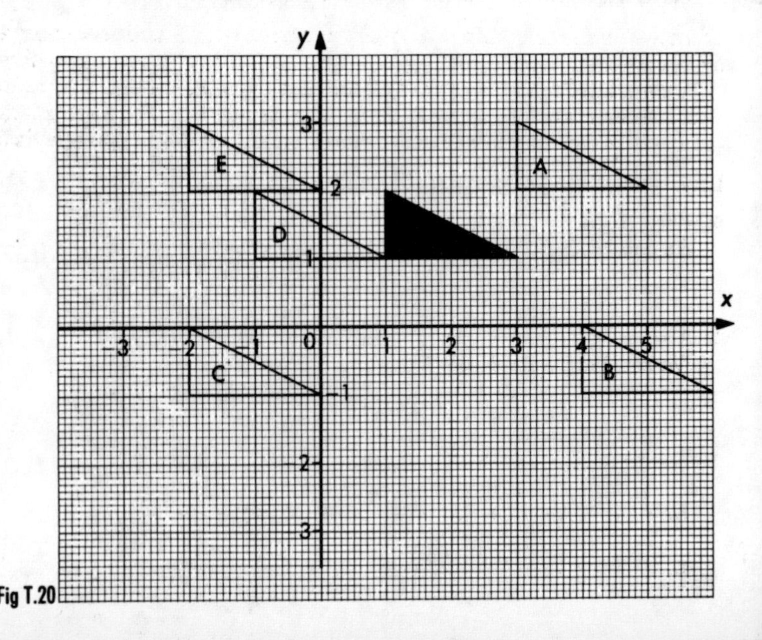

Fig T.20

TRANSLATIONS

A translation is a movement along the plane *without* any rotating, reflecting or enlarging. It is described by a movement horizontally and a movement vertically which we put together as a **vector.**

For example, the heavily shaded triangle in Figure T.20 has been *translated* to position A by moving 2 to the right and 1 up (notice how every point on and in the triangle moves in exactly the same way). We write this movement as a vector $\begin{pmatrix}2\\1\end{pmatrix}$

In a similar way we write the following translations:

- To B, 3 to the right and 2 down .... $\begin{pmatrix}3\\-2\end{pmatrix}$
- To C, 3 to the left and 2 down .... $\begin{pmatrix}-3\\-2\end{pmatrix}$
- To D, 2 to the left, nothing up .... $\begin{pmatrix}-2\\0\end{pmatrix}$
- To E, 3 to the left and 1 up .... $\begin{pmatrix}-3\\1\end{pmatrix}$

Notice how we used the *negative* to indicate movement *to the left* and to indicate movement *down.*

TRANSPOSITION

Transposition is the change of a subject in a formula. It is often necessary to change a formula round to help you to find a particular solution to a problem.

There is a 'Golden Rule' of transposition that you need to learn off by heart: 'If it's doing what it's doing to everything else on that side then you can move it to the other side of the equation and make it do the opposite thing'

Look through the following short examples to see this in action:

- $w = t + 5k$
  We could move the t, the $5k$ or the w to give $w - t = 5k$
  or $w - 5k = t$
  or $0 = t + 5k - w$

- $A = bh$
  We could move the b, the h or the A to give $A/b = h$
  or $A/h = b$
  or $1 = bh/A$

- $Y = \dfrac{t+2}{d}$
  We could move the (t+2), the d or the Y to give $\dfrac{Y}{t+2} = \dfrac{1}{d}$
  or $Yd = t + 2$ or $1 = \dfrac{t+2}{Yd}$

$p = m\ (x-3)$

We could move the m, the $(x-3)$ or the p to give

$\frac{p}{m} = x-3$ or $\frac{p}{x-3} = m$ or $1 = \frac{m\ (x-3)}{p}$

To fully transpose a formula though, is to make another letter the *subject* of that formula. This is often like sculpturing; in mathematics we need to mould our formula into what we want. This can be done as long as we remember to use our 'Golden Rule'.

See the following examples of this in action:

1 Make x the subject of the formula $y = 4\ (3x-1)$
We move things round until we finally end up with $x=$...
hence ... expand to give ... $y = 12x - 4$
then ... $y + 4 = 12x$
and $\frac{y+4}{12} = x$

2 Make t the subject of the formula $s = \frac{7+t^2}{3}$

We move things round until we finally end up with $t = ...$
hence ... $3s = 7 + t^2$
and ... $\sqrt{(3s-7)} = t$

TRANSVERSAL

Figure T.21 shows a pair of straight parallel lines and another straight line passing through them. This line passing through the parallel lines is called a *transversal*.

The angles created are shown as a and b and will be equal; they are called *alternate angles*. Both angles on the same side of the transversal are called *allied angles* and as such add up to 180°.

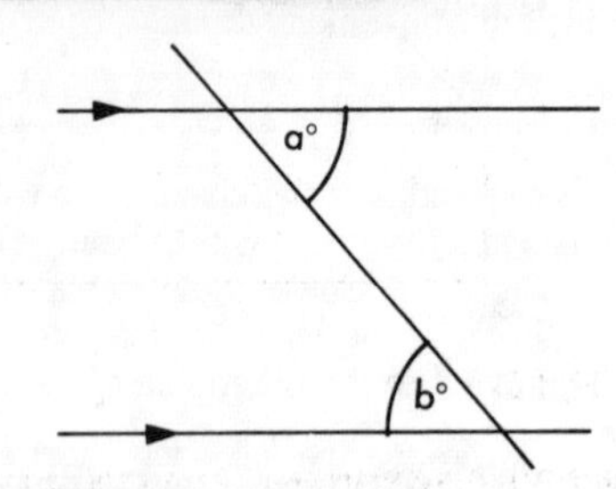

Fig T.21 Transversal

TRAPEZIUM

A trapezium is a **quadrilateral** that has a pair of sides parallel, as shown in Figure T.22.

The area of a trapezium is found by multiplying the height by the average length of the parallel sides. This is usually written as:

area $= \frac{h}{2}(a+b)$ or $h\ \frac{(a+b)}{2}$

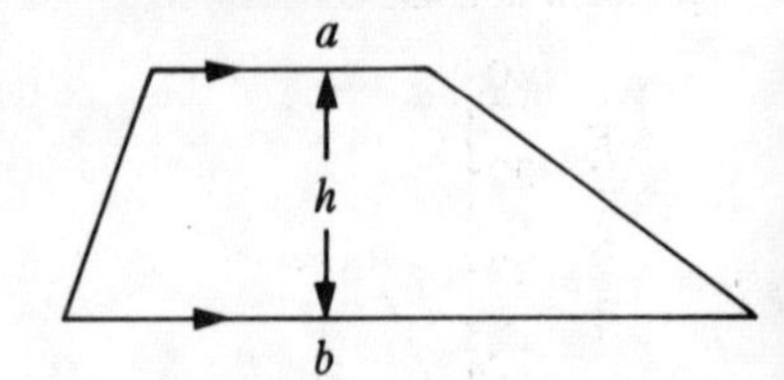

Fig T.22 Trapezium

where a and b are the lengths of the two parallel sides.

TRAPEZOIDAL METHOD (TRAPEZIUM RULE)

The trapezoidal method is a way of approximating the area under a curve by splitting that area up into trapeziums and finding the sum of all their areas added together.

For example, look at the curve in Figure T.23 and its area underneath. It has been split up into a number of trapeziums and it is a simple matter of calculating the area of each one and adding them together.

Note how the more trapeziums you split a shape into, the more *accurate* becomes your estimated area. (Find out all about this in A level mathematics!)

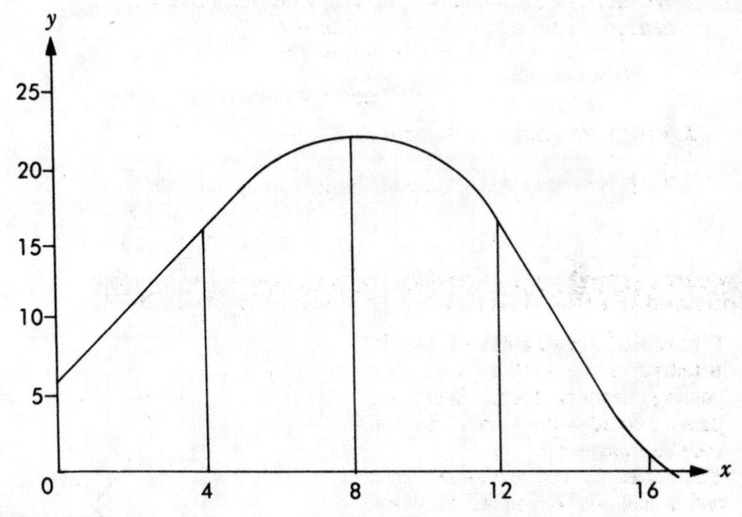

Fig T.23

TRAVEL GRAPHS

Travel graphs are distance/time graphs that represent a journey. Figure T.24 shows some travel graphs that illustrate particular situations. Note how the **gradient** of a line on a distance/time graph indicates the **speed** of travel.

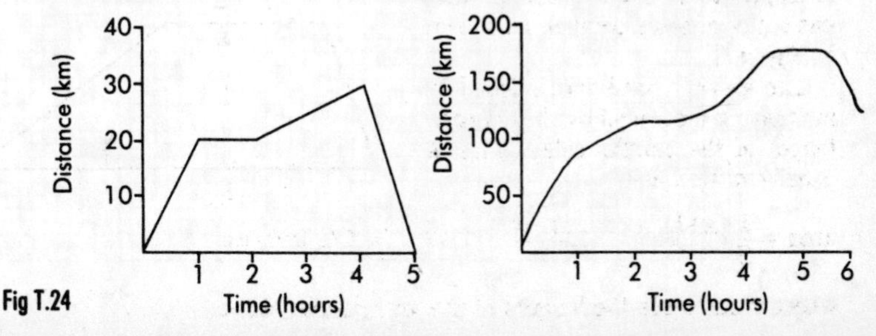

Fig T.24

Travel graphs often appear within a GCSE examination and ought to be read carefully and accurately. The biggest errors tend to be the misreading of scales or units on the axes.

■ Exam Questions

On a quiet Sunday morning, a police motorcycle parked in a layby was passed by a car. The police motor cyclist, suspecting that the car was exceeding the speed limit, set off after the car, overtook it and indicated that it should stop.

In Figure T.25, the straight line graph represents the car's journey over the first 30 seconds after passing the motor cycle.

a) What does the gradient of the straight line graph represent?
The police motor cycle's journey is indicated in the table below.

Time (seconds)	10	20	30	40	50	60
Distance (metres) from layby	180	460	880	980	1020	1020

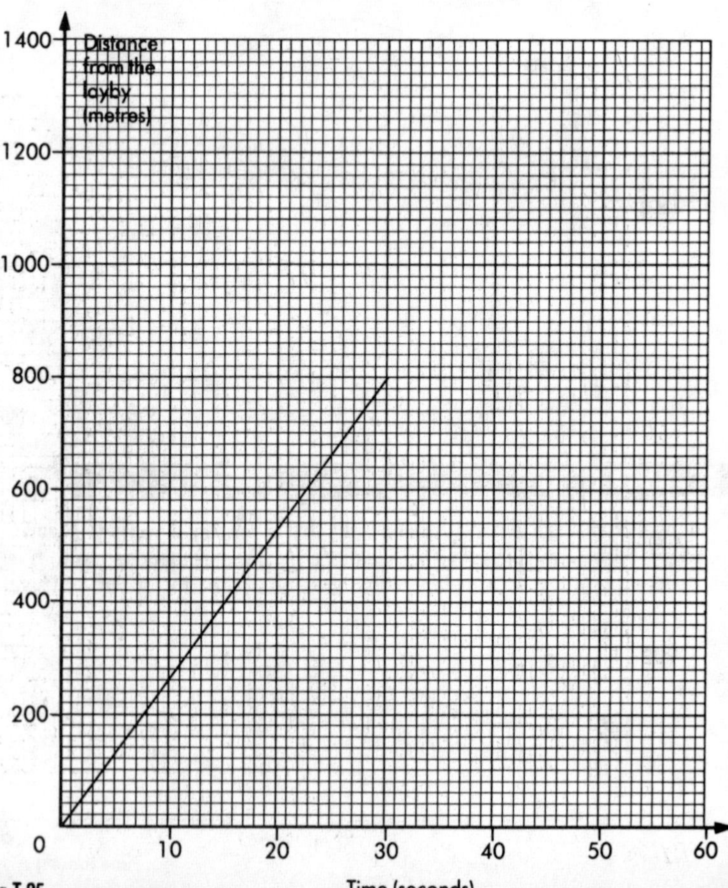

Fig T.25

b) On Figure T.25, plot the points representing the information in the above table and join them up with a smooth curve to represent the journey of the police motor cycle.
c) After how many seconds did the police motor cycle draw level with the car?
d) The car slowed down and stopped 20 metres behind the police motor cycle.
Complete the graph of the car's journey to illustrate the car slowing down and stopping.

(NEA; I)

■ Solution

a) The speed, or 1600m/minute, or 96 km/hour.
b) See Figure T.26.
c) 24 seconds, where the two lines cross.
d) See the dotted line on Figure T.26.

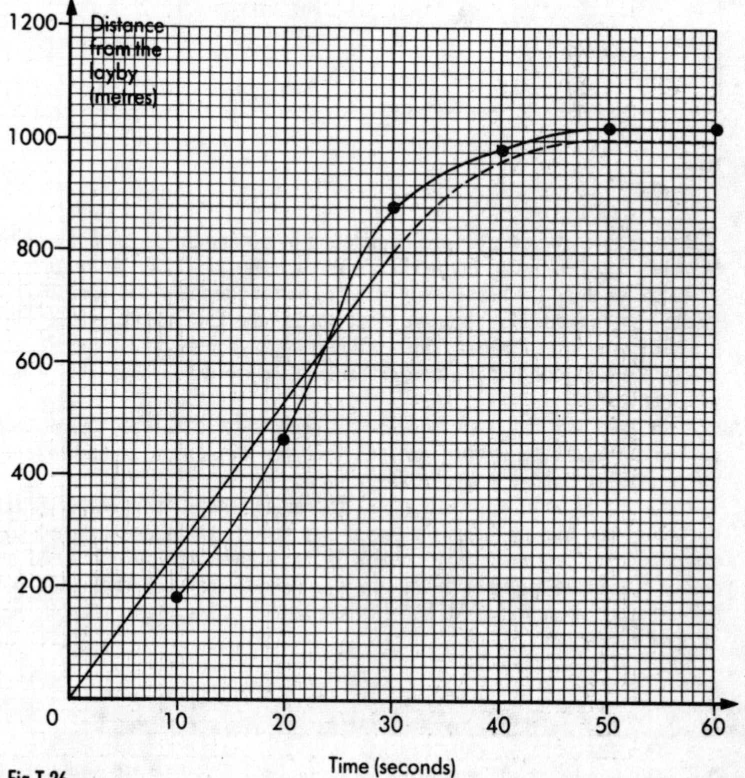

Fig T.26

TREE DIAGRAMS

A tree diagram is a particular way of illustrating all the events under consideration within a situation, with a view to calculating some probabilities. The diagram looks like branches off a tree and hence the name.

For example, the probability of Philip getting his maths homework correct is 0.4, while the probability of Tim's dad getting it correct is 0.8. What is the probability of only one of them getting the maths homework correct?

Philip	Tim's dad	Result	Probability
0.4 right	0.8 right	Both right	0.4 × 0.8 = 0.32
	0.2 wrong	Only Philip right	0.4 × 0.2 = 0.08
0.6 wrong	0.8 right	Only Tim's dad right	0.6 × 0.8 = 0.48
	0.2 wrong	Both wrong	0.6 × 0.2 = 0.12

Fig T.27

If we draw a tree diagram and complete it by putting the probabilities on the branches it will look like Figure T.27. We use the **AND** rule to *multiply* the probabilities along the branches in order to find the probabilities of each event in the end column. But then we need to use the **OR** rule to calculate the probability of either Philip or Tim's dad getting the homework correct. That is to say, we *add* together the two probabilities which will give us 0.08 + 0.48 which is 0.56.

If the question had been 'what is the probability of at least one of them getting the maths homework correct', then you would need to add together the probability 0.32 + 0.48 + 0.08 = 0.88.

The error to avoid while using tree diagrams is the obvious one of getting the adding and the multiplication routines mixed up. Or in fact, to use a tree diagram when it is not really helpful. If you know that you only need *one branch* of the tree diagram, then there is no need to draw the whole thing out, just the branch that you need. But do make sure that this really is all you need.

◄ Probability ►

TRIANGLES

A triangle is a plane shape with three straight sides. There are a number of special triangles that you ought to be familiar with.

TYPES OF TRIANGLE

Isosceles triangle

This has two of its sides the same and two angles the same, as indicated in Figure T.28. It has one **line of symmetry** bisecting the angle included between the two equal sides.

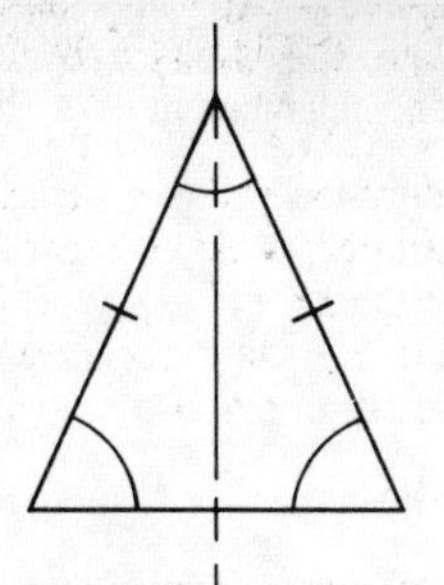

Fig T.28 Isosceles triangle

Equilateral triangle

This has all its sides the same length and all its angles are 60°. It has **three lines of symmetry** bisecting each angle, and its order of **rotational symmetry** is three (Fig T.29).

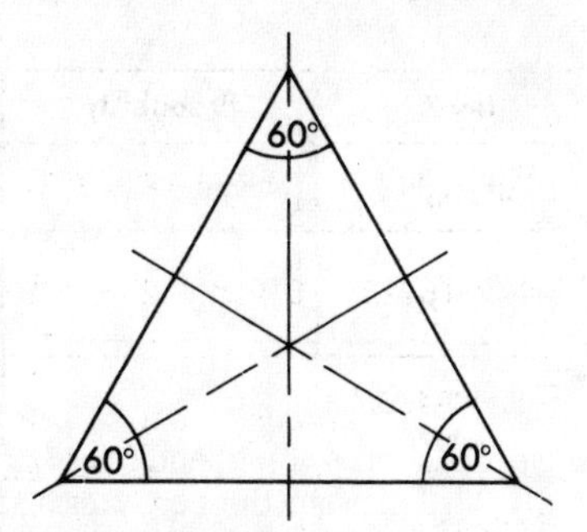

Fig T.29 Equilateral triangle

Right-angled triangle

This is one that contains a right angle (Fig T.30).

Scalene triangle

This is one which has all three sides a different length (Fig T.31).

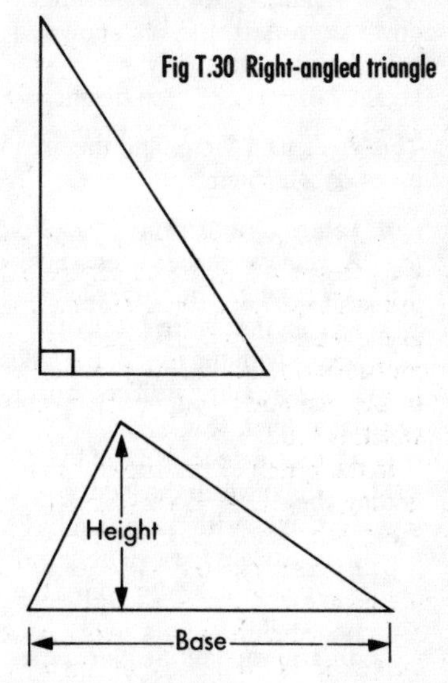

Fig T.30 Right-angled triangle

Fig T.32

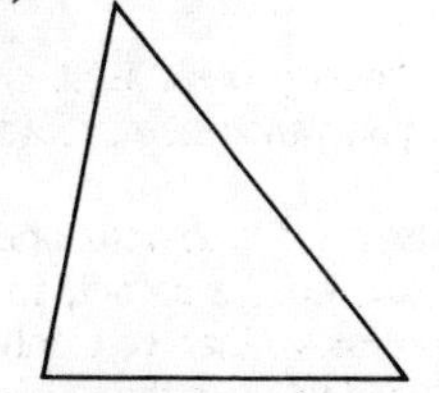

Fig T.31 Scalene triangle

AREAS OF TRIANGLES

The *area* of a triangle is calculated by multiplying half its base length by the perpendicular height, as in Figure T.32.

Area of triangle = ½ × base length × perpendicular height

If we know two sides and the included angle of a triangle, then we can calculate the area of a triangle by using trigonometry to calculate the perpendicular height. We then use the previous formula.

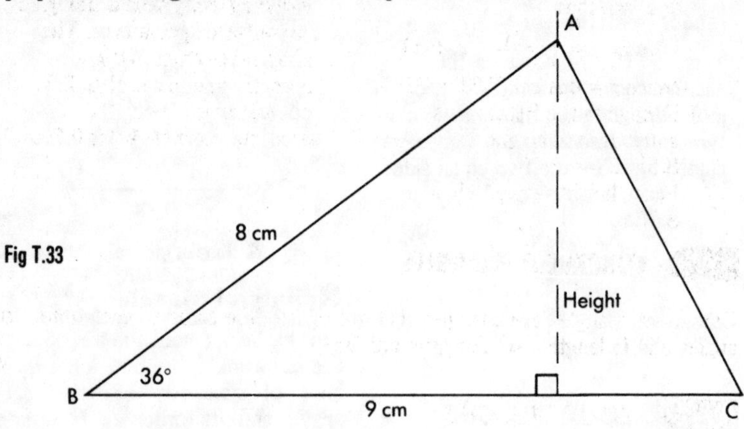

Fig T.33

For example, to find the area of triangle ABC, drop a perpendicular line down from A to BC, as shown in Figure T.33. Calculate the perpendicular height by trigonometry:
height / 8 = sin 36°, so height = 8 × sine 36°

This gives us 4.7 cm, and the area is now calculated by half of 4.7 × 9, which gives us 21.15 cm^2.

■ **Exam Questions**

A window cleaner uses a ladder 8.5 m long. The ladder leans against the vertical wall of a house with the foot of the ladder 2.0 m from the wall on horizontal ground. (Fig T.34).

a) Calculate the size of the angle which the ladder makes with the ground.
b) Calculate the height of the top of the ladder above the ground.
c) The window cleaner climbs 6.0 m up the ladder (see diagram). How far is his lower foot from the wall?

(MEG; I)

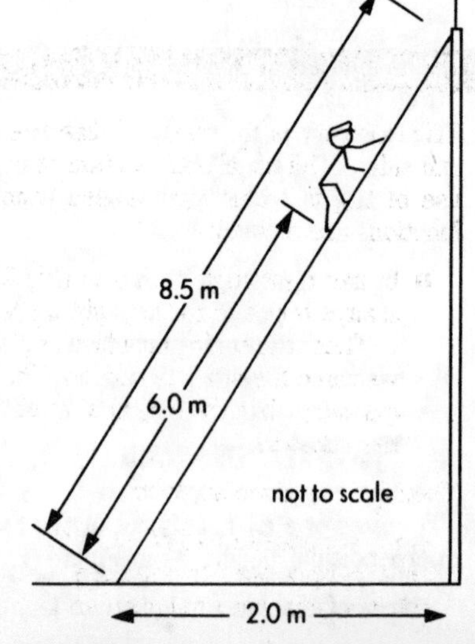

Fig T.34

■ **Solution**

a) Recognise the cosine situation, and that

$$\cos x = \frac{2}{8.5} = 0.2353$$

hence $x = \cos^{-1} 0.2353 = 76°$

b) Recognise the Pythagoras situation, and height^2 = $8.5^2 - 2^2 = 68.25$
hence height = $\sqrt{68.25}$ = 8.26m.

c) There are various methods including using similar triangles as well as trigonometry. The simplest perhaps is trigonometry and to calculate 2.5 cos 76° which (using the accurate value of 76°) is 0.59m.

CONGRUENT TRIANGLES

Congruent triangles are triangles that are exactly the same as each other, in angles and in length. ◀ Congruency ▶

SIMILAR TRIANGLES

Similar triangles are two triangles that are the same shape, that is their angles correspond to each other, but one is smaller than the other by some scale factor. In this case the ratios of each corresponding pair of sides is equal, and is the same as the scale factor of enlargement for the two triangles. ◀ Similarity ▶

TRIGONOMETRICAL GRAPHS

◀ Graphs ▶

TRIGONOMETRY

Trigonometry is the study of features of triangles, mainly to do with angles and sides. The usual abbreviation of trigonometry is trig. The most common use of trig is within right-angled triangles, and it is in these that our trig functions are defined:

■ In any right-angled triangle (Fig T.35) we call the *longest side*, which is always opposite to the right angle, the *hypotenuse*.

Then, depending on which angle of the triangle we are finding or using, we name the other two sides. The side opposite to the angle is called the *opposite*; while the side next to both the angle and the right angle is called the *adjacent*.

Then for any given angle x:

tan x = opposite / adjacent
sin x = opposite / hypotenuse
cos x = adjacent / hypotenuse

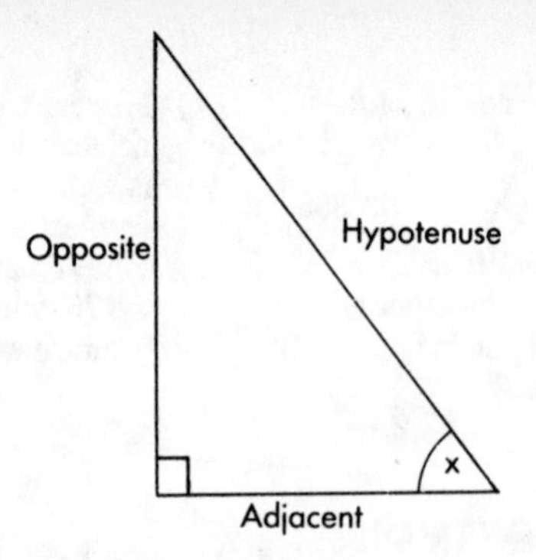

Fig T.35

This information is likely to be given on a formula sheet within an examination, but if the facts are known then it makes problem solving much quicker.

One way of remembering the facts is to remember a sentence such as 'Tommy On A Ship Of His Caught All Herring' taking the first letter of each word gives us T=O/A, S=O/H, C=A/H. There are lots of other similar sentences to be made like this to help you to remember the trig fractions. Make up your own!

Note: for the use of tan, sin and cos, see their individual entry in this reference guide.

With this topic there are quite a few mistakes that are made:

- Using the wrong trig function. Do look carefully at which angle you're given or need to find and then work out. Is it tan, sin or cos?
- Often the fraction is written upside down, which will cause problems. So get it the right way up, and notice that sin and cos *must* be a fraction smaller than 1.
- Candidates often make mistakes because they have not got a clear diagram of the situation. So they either calculate the wrong angle altogether or just use the wrong data. Start with a *clear diagram* of the problem given, especially if the problem has come from a three-dimensional situation.

■ Exam Questions

Figure T.36 shows two stages of a cable car route.
Safety experts have decided that a cable car stage is only safe if the angle of elevation is less than 68° and greater than 30°.

a) Calculate the angles of elevation, x and y, indicated on the diagram.
   i) x
   ii) y
b) Based on the information given, is each stage shown of this route safe? Give reasons for your answers.

(NEA; I)

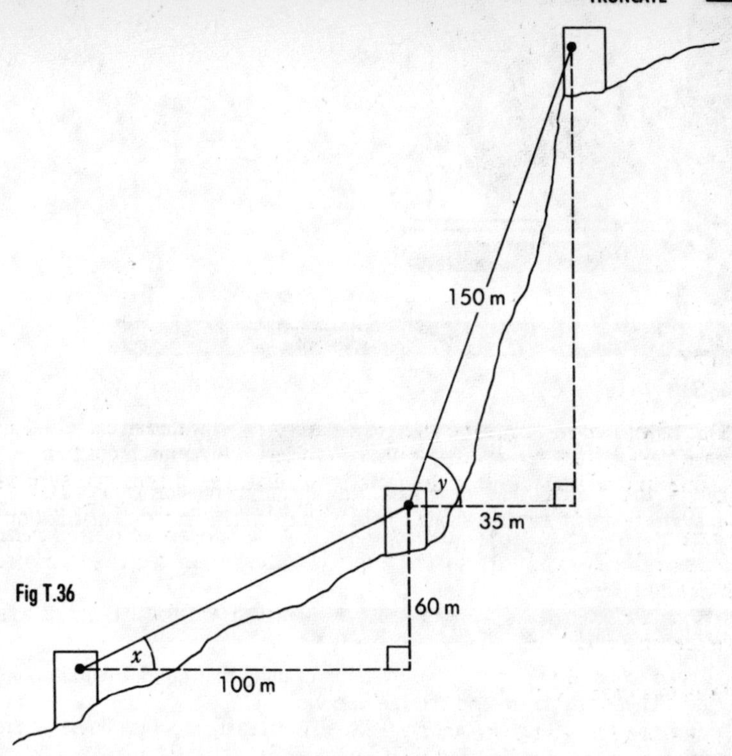

Fig T.36

- **Solution**

a) i) Recognise the tan situation to give $\tan x = 60/100 = 0.6$
hence $x = \tan^{-1}0.6 = 31°$
ii) Recognise the cos situation to give $\cos y = 35/150 = 0.2333$
hence $y = \cos^{-1}0.2333 = 76.5°$ or $77°$

b) The first stage is safe since $x = 31°$ and this is between the acceptable angles. Yet the second stage is not safe since the angle $y = 77°$ is greater than the highest safe angle of 68°.

TRUNCATE

This means to cut off, or shorten. In mathematics, it means to remove the decimal fraction.

UNION

◀ Sets ▶

UNIT MATRIX

This is the matrix that describes the identify transformation, i.e. a transformation that leaves a shape where it started. The unit 2×2 matrix (for 2 rows and 2 columns) is $\begin{pmatrix} 1 & 0 \\ 0 & 1 \end{pmatrix}$

UNITS

You need to be aware of the two systems of units that operate within our society. These are the **metric** and the **imperial** systems.

THE METRIC SYSTEM

Examples of metric system units are:

1 kilogram = 1000 grams
1 kilometre = 1000 metres
1 kilowatt = 1000 watts

It is helpful to remember that the word 'kilo' means 1000.

1000 kilograms = 1 tonne
10 millimetres = 1 centimetre
100 centimetres = 1 metre
1000 millimetres = 1 metre
1000 millilitres = 1 litre

These ought to be learnt.

THE IMPERIAL SYSTEM

Examples of Imperial system units are:

12 inches = 1 foot

3 feet = 1 yard
16 ounces = 1 pound
8 pints = 1 gallon

There are plenty more Imperial units. However, these are the more common ones that you would be expected to be familiar with, although you would not be expected to learn them.

EQUIVALENCE OF UNITS

You ought to be aware of the approximate *conversion* from the popular *imperial units* to the *metric*:

2 pounds weight is approximately 1 kilogram
3 feet is approximately 1 metre
5 miles is approximately 8 kilometres
1 gallon is approximately 4.5 litres

Although you will find conversion questions in some GCSE examination papers, the conversion factor will usually be given. The biggest error associated with units, is the *lack* of them in an answer or the use of the *wrong ones*.

When a question is asked, then the answer (if it has units like cm, kg, mph, etc) should be given in the *correct* units. You could lose marks if no units are given or the wrong ones are used. So always *check* that the units you have used are indeed the ones used in the question, unless you have needed to convert them to a more appropriate unit.

For example, what is the length of a street that is equivalent to 150 paving slabs end-to-end, where the slabs are each 75cm long?

We simply multiply 75cm by 150. This comes to 11,250cm. Yet it would be more sensible to give the answer as 112.5 metres.

UNIVERSAL SET

The universal set is the **set** that defines the limit of your situation. In other words, it states the *only* things you are talking about.

For example, if your universal set is **integers**, then the only numbers you can consider are whole ones and no fractions.

The shorthand for universal set is $\mathscr{E}$ and is also usually placed on the outside of a **Venn** diagram to signify that this is the whole situation.

Where $\mathscr{E}$ is the universal set containing the set A, then:

$\mathscr{E} \cap A = A$
$\mathscr{E} \cup A = \mathscr{E}$

VARIABLE

A variable is something that can change. People's moods are variable, since they can change from day-to-day and even minute-to-minute.

In mathematics a variable is usually a letter in an algebraic expression that stands for something that can change. For example, the formula for the area of a rectangle is area = length × base, i.e. $A = l \times b$ where A, l and b are all variables as they can all be different values depending on the particular rectangle.

Variables are commonly used in computer programs in just the same way. For example, you can define a *loop* or an expression with a variable:

```
FOR I = 1 TO 15
```

This is the start of a computer program loop where the I will vary from 1 to 15 as the program loops round.

VARIATION

Variation is the study of how one thing is in proportion to another. ◀ Proportion ▶

VAT

VAT is the word that means 'Value Added Tax'. It is a tax that the Government places on goods in order to help raise revenue from its people. It is a percentage rate; at the moment this rate is 15% on most goods, but there are a few categories where this rate is zero, for example children's clothes and books.

For example, if a television is sold at a price of £450, including the 15% VAT, how much money will go to the government as VAT?

The final price included the 15% VAT and therefore represented 115% of the pre-tax price. Hence the pre-tax price was 450 × 100 / 115 which is £391.30, so the VAT is the difference of £58.70 ◀ Taxes ▶

VECTORS

A vector is a column matrix. The scientists will properly define it as a force having direction and magnitude, yet the use we put it to in GCSE Mathematics is not quite as rigid as that.

In mathematics a vector can be thought of as a *displacement*, a movement from one place to another. This displacement is defined by its horizontal and vertical movement.

VECTOR NOTATION

There are three common ways to write a vector:

1. Write it as $\overrightarrow{AB}$ where AB is some given line and the direction of the vector is from A to B.
2. Label it as $\underline{a}$ or $\underset{\sim}{a}$ to denote that this is a vector; there could well be an arrow on the line to show the direction of the vector.
3. It can simply be defined as the column matrix $\begin{pmatrix} a \\ b \end{pmatrix}$ where a is the movement horizontally to the right (so a negative value will indicate movement to the left), and b is the vertical movement up (so a negative value will indicate a movement down). See the examples in Figure V.1.

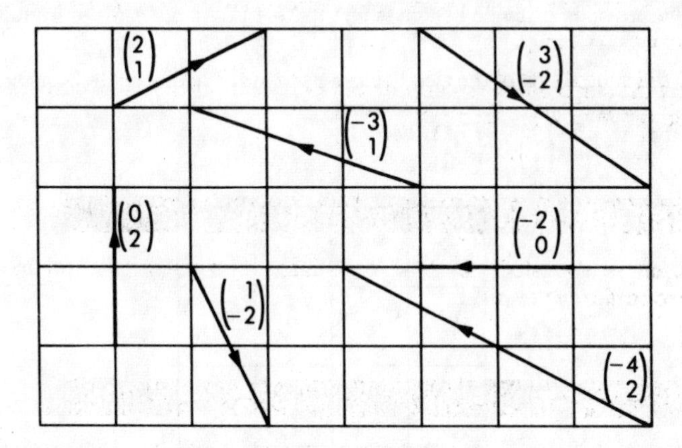

Fig V.1

Vector addition

We *add* vectors together as the sum of their movements. Notice how in Figure V.2 that the sum of the two vectors $\underline{a}$ and $\underline{b}$ are represented by one vector $\underline{a} + \underline{b}$, that is by the sum of its horizontal movements and of its vertical movements.

In other words: $\begin{pmatrix}3\\1\end{pmatrix} + \begin{pmatrix}4\\-3\end{pmatrix} = \begin{pmatrix}7\\-2\end{pmatrix}$

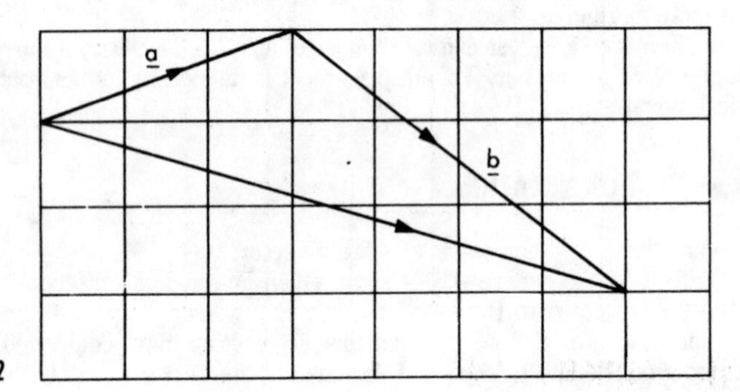

Fig V.2

Vector subtraction

We *subtract* two vectors by adding the *negative* of the vector to be subtracted.

So the previous example becomes $\begin{pmatrix}3\\1\end{pmatrix} + \begin{pmatrix}-4\\3\end{pmatrix}$. As we can see from Figure V.3, the subtraction of the two vectors used in the addition $\underline{a} + \underline{b}$ can be seen as the subtraction of their movements.
In other words: $\begin{pmatrix}3\\1\end{pmatrix} - \begin{pmatrix}4\\-3\end{pmatrix} = \begin{pmatrix}-1\\4\end{pmatrix}$

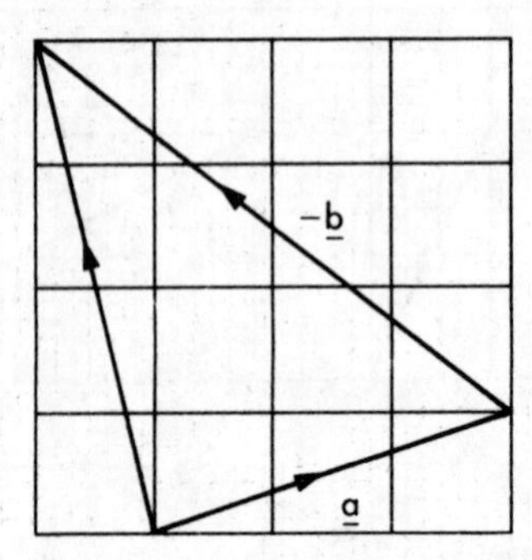

Fig V.3

Vector multiplication

We *multiply* vectors by repeated addition. That is to say $3\underline{a} = \underline{a} + \underline{a} + \underline{a}$ as seen in Figure V.4. Also ½ $\underline{a}$ will be half the vector $\underline{a}$, as also seen on the figure.

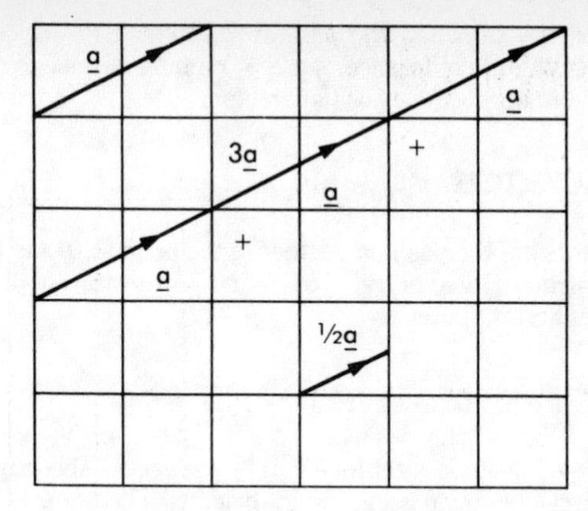

Fig V.4

POSITION VECTORS

The position vector of a point is the column vector from a defined origin to that point. For example the position vector of the point D in Figure V.5 is $\begin{pmatrix} 3 \\ 4 \end{pmatrix}$.

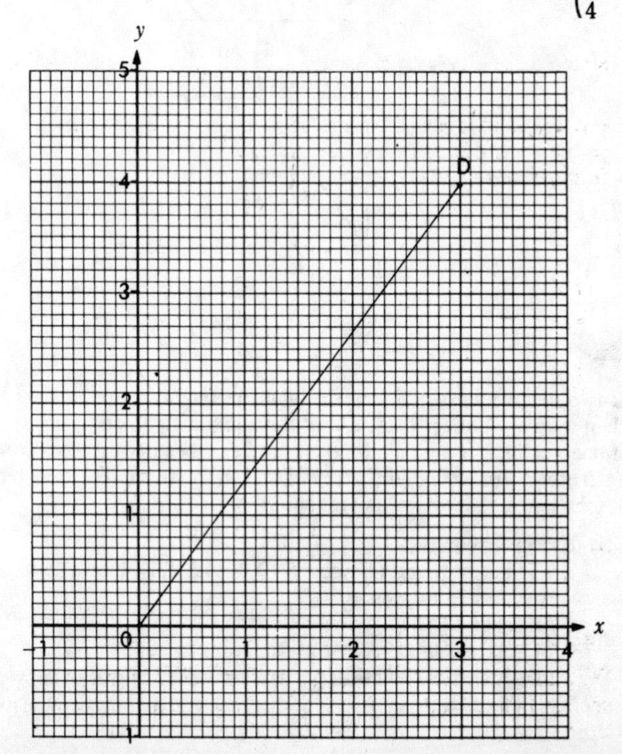

Fig V.5

Note the similarity between the *position vector* and the *co-ordinate* of D. However, also note the difference, since a common mistake is to write down the co-ordinate instead of the position vector.

BASE VECTORS

The base vectors are the position vectors to the point (1,0) and (0,1), as used in transformation geometry to find **transformation matrices.**
◄ Transformation matrices ►

MAGNITUDE OF A VECTOR

The magnitude of a vector is represented by its length. The magnitude of any column vector can be found using the Pythagoras' Theorem. For example, in Figure V.6 the magnitude of the vector $\begin{pmatrix}3\\4\end{pmatrix}$ is found, using Pythagoras, as 5.

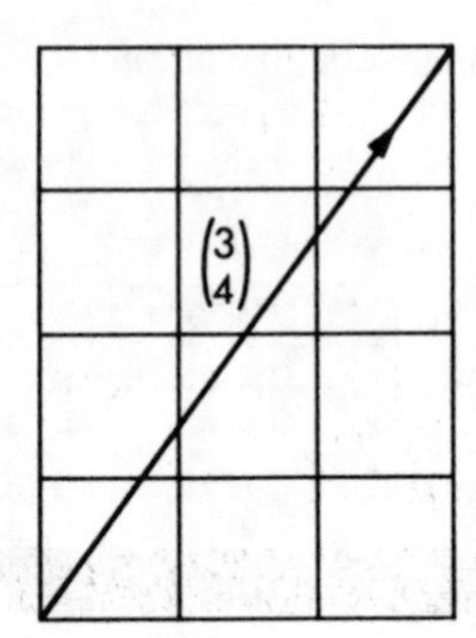

Fig V.6

PARALLEL VECTORS

If two vectors have the same column vector or are defined as being equal, then they are both parallel and of the same magnitude.
If one vector is a multiple of another, then those two vectors are parallel to each other. For example, the vectors $\underline{a}$ and $4\underline{a}$ are parallel to each other and $4\underline{a}$ is four times as long (or as big) as $\underline{a}$.

- **Exam Questions**

1 The line segment *OP* joins *O* (0,0) to *P*(1,1). Find a matrix of the transformation which
  a) doubles the length of *OP*,
  b) rotates *OP* clockwise about *O* through 90°,
  c) doubles the length of *OP* and rotates *OP* through 90° clockwise about *O*.

(NEA; H)

■ 2 In the triangle shown in Figure V.7, $\overrightarrow{AB} = \mathbf{p}$ and $\overrightarrow{AC} = \mathbf{q}$.

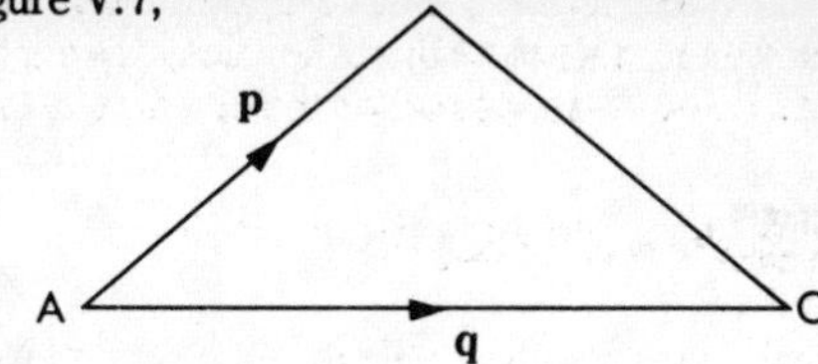

Fig V.7

a) Express $\overrightarrow{BC}$ in terms of **p** and **q**

b) If M and N are the mid-points of AB and AC respectively, express in terms of **p** and **q**

i) $\overrightarrow{MB}$,

ii) $\overrightarrow{MN}$.

c) What **two** facts can you deduce about the line MN from the result obtained in b) part ii)?

(NEA; H)

■ **Solutions**

1 a) $\begin{pmatrix} 2 & 0 \\ 0 & 2 \end{pmatrix}$ b) $\begin{pmatrix} 0 & 1 \\ -1 & 0 \end{pmatrix}$ c) $\begin{pmatrix} 0 & 2 \\ -2 & 0 \end{pmatrix}$

The simplest way was to have considered the base vectors of $\begin{pmatrix} 1 & 0 \\ 0 & 1 \end{pmatrix}$ and determine their image.

2 a) $\overrightarrow{BC} = \mathbf{q} - \mathbf{p}$

b) i) $\overrightarrow{MB} = \frac{1}{2}\mathbf{p}$ ii) $\overrightarrow{MN} \frac{1}{2}\mathbf{q} - \frac{1}{2}\mathbf{p} = \frac{1}{2}(\mathbf{q} - \mathbf{p})$

c) MN is parallel to BC and,
MN is half the length of BC.

VENN DIAGRAM

A Venn Diagram is one that illustrates **sets**. It is often clearer to illustrate a situation on a Venn Diagram than to use words in an explanation. Figure V.8 shows some Venn Diagrams. ◀ **Set notation** ▶

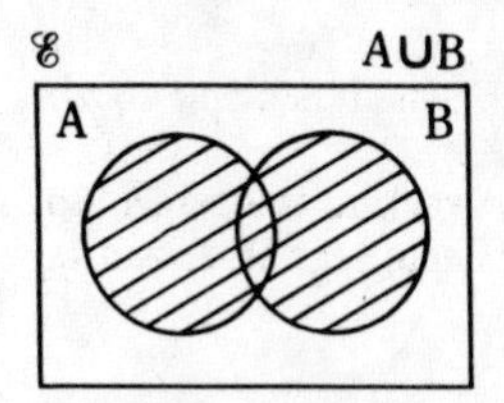

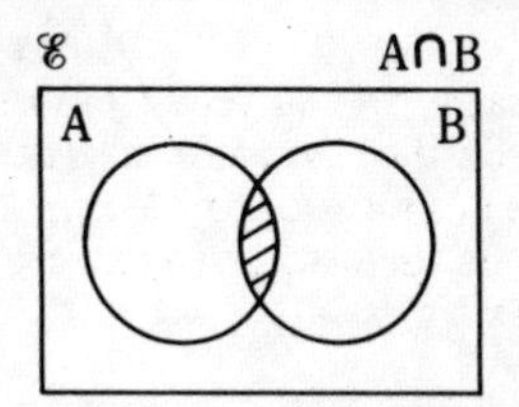

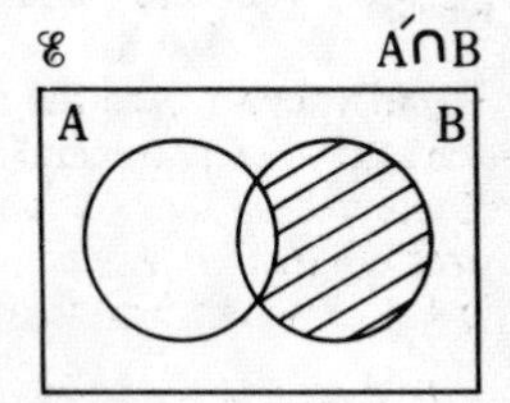

Fig V.8 Venn diagrams

VERTEX

A vertex is a point where two lines or two edges meet. The plural of vertex is *vertices*. The vertices of plane shapes are the sharp corners where the angles are found. The vertex of a pyramid is the top where all the edges meet. (Fig V.9).

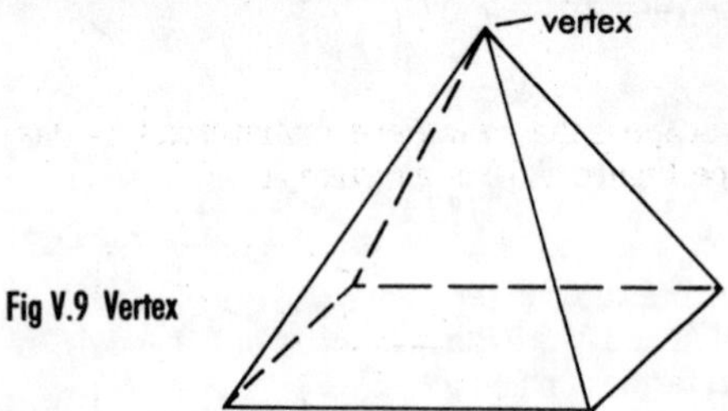

Fig V.9 Vertex

■ **Exam Questions**

Figure V.10). shows the net for a solid.
On the diagram mark with an X the other point or points that will meet the point A to form a vertex when the solid is made.

(NEA; I)

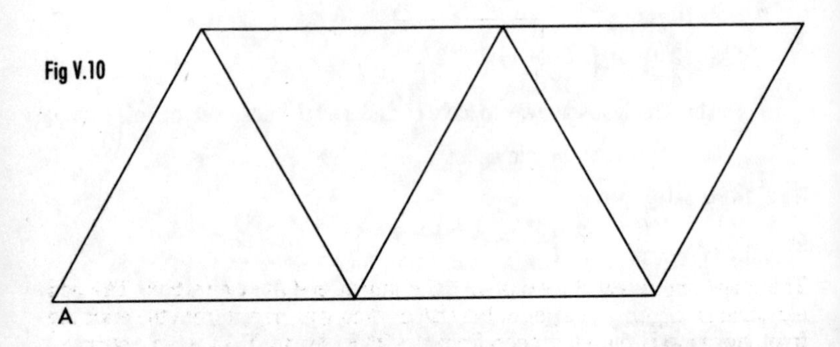

Fig V.10

■ **Solution**

See Figure V.11.

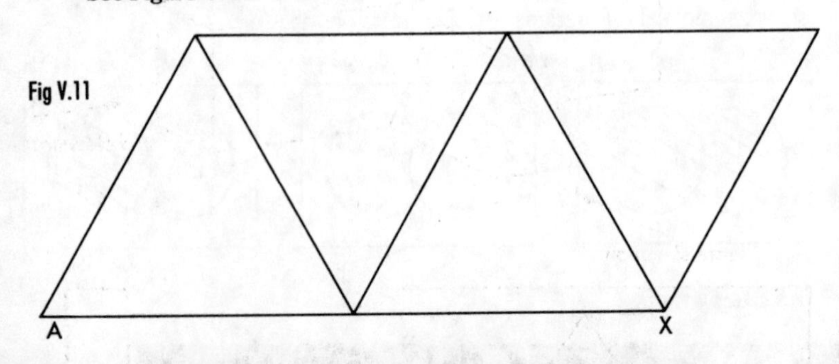

Fig V.11

VIEWS

When looking at solid shapes you get different views depending on which direction you look at the shape. The two important views are the plans and the elevations.

Plans

The plan of a shape is the view you get when you look directly overhead down onto the shape. Figure V.12 is an example.

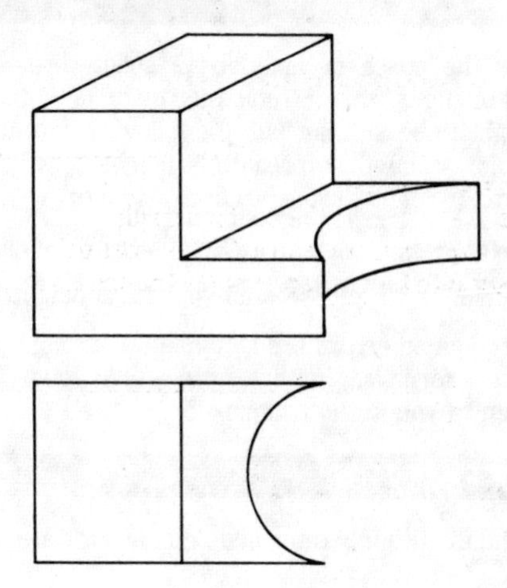

Fig V.12 View; plan

Elevation

There are two types of elevation, *end elevation* and *front elevation*. The end elevation is the view you get of the shape when you view it from the end; the front elevation is the view seen from the front. Figure V.13 is an example.

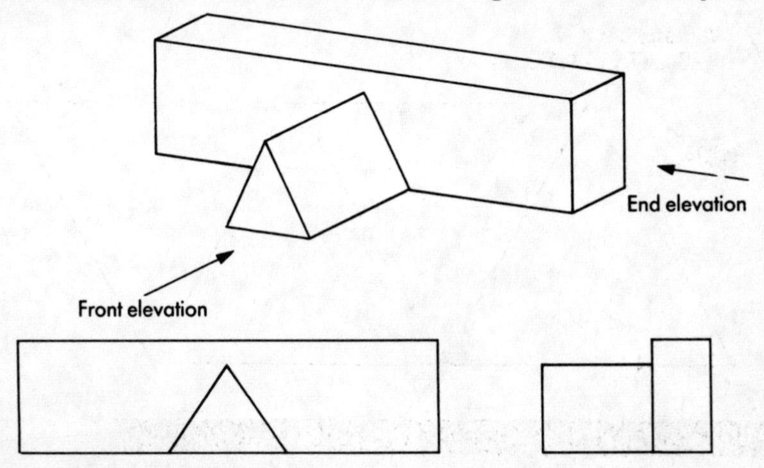

Fig V.13 View; elevation

The most common error made with views is to try and make them look three-dimensional instead of *two-dimensional*. You must not try to put any *perspective* into the diagram; nor should you draw what you *think* you would see (because we see in three dimensions, and not in 2 dimensions). Draw the shape of the top, or the end, or the front!

VOLUME

Volume is the space occupied by a solid, three-dimensional shape. It is measured in cubes, that is cubic metres or m^3, etc.

You ought to be familiar with the following formulae for finding volumes:

- Cuboid: = length × breadth × height
- Prism: = regular cross section × length
- Sphere: = $\frac{4\ \pi r^3}{3}$
- Pyramid: = ⅓ × base area × vertical height

The most common error here is to use the wrong formula. It is helpful to you to learn these formulae, but if you can't, then you will find them on the formula sheet given to you in the examination.

VULGAR FRACTION

This is when both numerator and denominator are whole numbers, e.g. ¾.

WAGES

Wages are the payments made to employees by the employers. That is to say, the pay given to the workers by those who hired them. Wages are usually defined as weekly payments and can often vary with the number of hours worked or the number of articles made or jobs done.

If someone has a basic working week of 40 hours, then any hours extra to this time is called *overtime*. Overtime is usually paid by various rates:

- time and a quarter — is basic hourly rate × 1.25
- time and a half — is basic hourly rate × 1.5
- double time — is basic hourly rate × 2

For example, John worked a 43-hour week, where his basic week is 38 hours at a rate of £4.50 per hour. His overtime rate is time and a quarter. What is his week's wage?

The basic pay is £4.50 × 38 = £171
The overtime pay is £4.50 × (43−38) × 1.25 = £28.12
Hence the total pay is £199.12

Of course tax would probably have to be paid on this wage but the question has not asked us to calculate that here.

- **Exam Questions**

Maria earns £2.20 for each hour she works as a shop assistant. She works from 9 a.m. to 1 p.m. and from 2 p.m. to 6 p.m. each day for five days a week.

a) Calculate Maria's weekly wage.

b) Maria is asked to work four hours' overtime on her day off. She is paid twice the rate of £2.20 per hour. What is Maria paid for working the four hours' overtime?

(NEA; L)

- **Solution**

a) 8 hours each day. Hence weekly wage will be
8×5×£2.20 = £88

b) Overtime rate = £2.20×2 = £4.40.
Hence overtime pay = £4.40×4 = £17.60

WEIGHING SCALES

Weighing scales come in all sorts of shapes and sizes, yet in most cases the scale we have to read is similar to the ones shown. You must be able to read quite accurately the readings on weighing scales, as well as on the other types of scales you will come across.

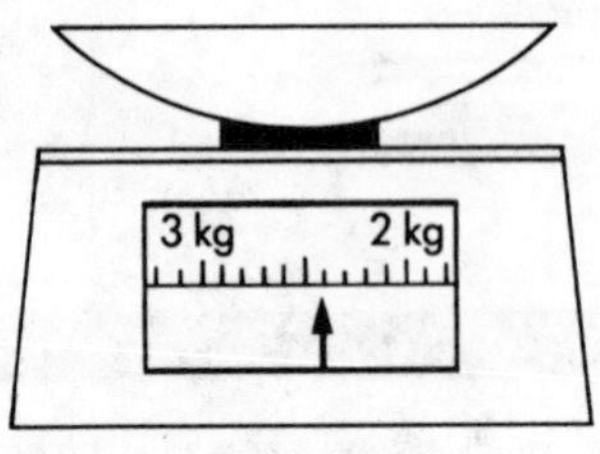

Fig W.1

Figure W.1 shows a set of weighing scales; notice how the scale reads from right to left. Between each kilogram the space is divided into ten parts, and as one tenth of 1 kg is 100 grams, each small line represents 100 g. You can see that the pointer is on the fourth line between the 2 kg and 3 kg marks so the object we are weighing is 2 kg 400 g, or 2.4 kg.

Notice how on the weighing scales in Figure W.2 the space between each kilogram is divided into five parts, each line representing one-fifth of a kilogram which is 200 grams. Hence the pointer is pointing to 3 kg 600 g or 3.6 kg.

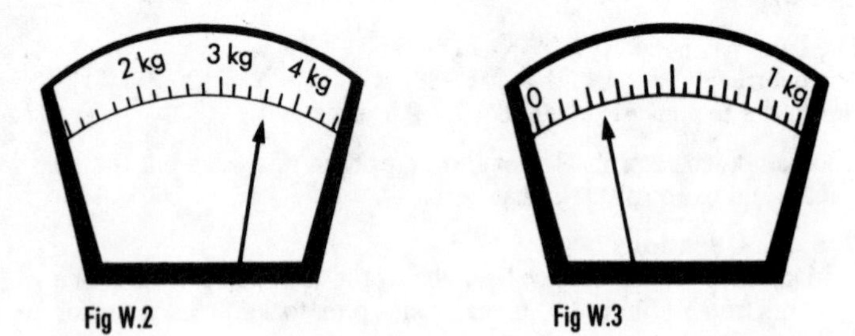

Fig W.2

Fig W.3

Now look at the weighing scales in Figure W.3. The space between the kilogram is divided into ten large parts (longer lines), each one now representing 100 g (or 0.10 kg), and each of these spaces is divided into two parts, each one 50g (0.05 kg). The pointer on the diagram is pointing to 0.25 kg.